Lecture Notes in Computer

Edited by G. Goos, J. Hartmanis and J.

Springer

Berlin
Heidelberg
New York
Barcelona
Hong Kong
London
Milan
Paris
Singapore
Tokyo

Jean-Marc Champarnaud   Denis Maurel
Djelloul Ziadi (Eds.)

Automata Implementation

Third International Workshop
on Implementing Automata, WIA'98
Rouen, France, September 17-19, 1998
Revised Papers

Springer

Series Editors

Gerhard Goos, Karlsruhe University, Germany
Juris Hartmanis, Cornell University, NY, USA
Jan van Leeuwen, Utrecht University, The Netherlands

Volume Editors

Jean-Marc Champarnaud
Djelloul Ziadi
University of Rouen, Computer Science Laboratory
F-76821 Mont-Saint-Aignan Cedex, France
E-mail: {jmc,dz}@dir.univ-rouen.fr

Denis Maurel
LI/E3i University of Tours
64 avenue Jean Portalis, F-37200 Tours, France
E-mail: maurel@univ-tours.fr

Cataloging-in-Publication data applied for

Die Deutsche Bibliothek - CIP-Einheitsaufnahme

Automata implementation : revised papers / Third International Workshop on
Implementing Automata, WIA '98, Rouen, France, September 17 - 19, 1998.
Jean-Marc Champarnaud ... (ed.). - Berlin ; Heidelberg ; New York ; Barcelona ;
Hong Kong ; London ; Milan ; Paris ; Singapore ; Tokyo : Springer, 1999
 (Lecture notes in computer science ; Vol. 1660)
 ISBN 3-540-66652-4

CR Subject Classification (1998): F.1.1, F.4.3, I.2.7, I.2.3, I.5, B.7.1

ISSN 0302-9743
ISBN 3-540-66652-4 Springer-Verlag Berlin Heidelberg New York

© Springer-Verlag Berlin Heidelberg 1999
Printed in Germany

Typesetting: Camera-ready by author
SPIN: 10704143   06/3142 – 5 4 3 2 1 0   Printed on acid-free paper

Foreword

The papers contained in this volume were presented at the third international Workshop on Implementing Automata, held September 17–19, 1998, at the University of Rouen, France.

Automata theory is the cornerstone of computer science theory. While there is much practical experience with using automata, this work covers diverse areas, including parsing, computational linguistics, speech recognition, text searching, device controllers, distributed systems, and protocol analysis. Consequently, techniques that have been discovered in one area may not be known in another. In addition, there is a growing number of symbolic manipulation environments designed to assist researchers in experimenting with and teaching on automata and their implementation; examples include FLAP, FADELA, AMORE, Fire-Lite, Automate, AGL, Turing's World, FinITE, INR, and Grail. Developers of such systems have not had a forum in which to expose and compare their work. The purpose of this workshop was to bring together members of the academic, research, and industrial communities with an interest in implementing automata, to demonstrate their work and to explain the problems they have been solving.

These workshops started in 1996 and 1997 at the University of Western Ontario, London, Ontario, Canada, prompted by Derick Wood and Sheng Yu. The major motivation for starting these workshops was that there had been no single forum in which automata-implementation issues had been discussed. The interest shown in the first and second workshops demonstrated that there was a need for such a forum. The participation at the third workshop was very interesting: we counted sixty-three registrations, four continents, ten countries, twenty-three universities, and three companies.

The general organization and orientation of WIA conferences is governed by a steering committee composed of Jean-Marc Champarnaud, Stuart Margolis, Denis Maurel, and Sheng Yu, with Derick Wood as chair. The WIA 1999 meeting will be held at the University of Potsdam, Germany, and the 2000 meeting in London, Ontario.

June 1999   Jean-Marc Champarnaud Denis Maurel Djelloul Ziadi

Organization

WIA'98 was organized in France by Jean-Marc Champarnaud and Djelloul Ziadi, LIFAR, University of Rouen, and Denis Maurel, LI/E3i, University of Tours.

Executive Committee

Conference Co-chairs:	Jean-Marc Champarnaud
	Denis Maurel
Program Committee Co-chairs:	Jean-Marc Champarnaud
	Denis Maurel
	Djelloul Ziadi
Conference Coordinators:	Pascal Caron
	Jean-Luc Ponty
	Djelloul Ziadi

Program Committee

Anne Bruggemann-Klein	Technische Universität Munchen, Germany
Jean-Marc Champarnaud	Université de Rouen, France
Max Garzon	University of Memphis, Tennessee, USA
Franz Gunthner	Ludwig Maximillian Universität, Germany
Nicolas Halbwachs	CNRS, VERIMAG, France
Helmut Jurgensen	University of Western Ontario, Canada
	Universität Postdam, Germany
Stuart Margolis	Bar Ilan University, Israel
Denis Maurel	Université de Tours, France
Mehryar Mohri	AT&T Labs-Research, USA
Gene Myers	University of Arizona, Tucson, USA
Jean-Eric Pin	CNRS, Université Paris 7, France
Darrell Raymond	Gateway Group Inc., Canada
Emmanuel Roche	Teragram Corporation, Boston, USA
Susan Rodger	Duke University, USA
Kai Salomaa	University of Western Ontario, Canada
Jorge Stolfi	Universidade de Campinas, Brazil
Wolfgang Thomas	Universität Kiel, Germany
Bruce Watson	Ribbit Software Systems Inc., Canada
Derick Wood	HK University of Science & Technology, Hong-Kong
Sheng Yu	University of Western Ontario, Canada
Djelloul Ziadi	Université de Rouen, France

Invited Talk

We thank Professors Georges Hansel and Maurice Nivat for their very interesting talks (that are not included in this book):

Georges Hansel   *Automata and Logic*
Maurice Nivat   *Equivalences of NDFA's and Monoidal Morphisms*

Sponsoring Institutions

We would like to warmly thank all the sponsors who agreed to support WIA'98

Our administrative authorities were particularly generous:

- The Faculty of Science and Technology at Rouen
- The Scientific Council of the University
- The Ministry of National Education Research and Technology
- The Research Program PRC/GDR AMI (which involves the Ministry and the C.N.R.S.)

The local authorities also agreed to participate financially. Let us thank today:

- The Ville de Mont-Saint-Aignan which took in charge the printing of the pre-proceedings
- The Ville de Rouen
- The Conseil Général de Seine-Maritime
- The Conseil Régional de Haute-Normandie

A special thanks goes to the companies and their managers who paid particular attention to WIA topics. Let us mention here:

- Micro-Technique Rouen (micro-computer distributor)
- Hewlett-Packard and Medasys-Digital-Systems (distributor)
- TRT Lucent Technologies (telephony, embedded software)
- EDF Division Recherche
- Matra Systèmes et Information
- Dassault Aviation
- Cap Gemini

Table of Contents

Extended Context-Free Grammars and Normal Form Algorithms*

Jürgen Albert[1], Dora Giammarresi[2], and Derick Wood[3]

[1] Lehrstuhl für Informatik II
Universität Würzburg
Am Hubland, D-97074 Würzburg, Germany
albert@informatik.uni-wuerzburg.de
[2] Dipartimento di Matematica Applicata e Informatica
Università Ca' Foscari di Venezia
via Torino 155, 30173 Venezia Mestre, Italy
dora@dsi.unive.it.
[3] Department of Computer Science
Hong Kong University of Science & Technology
Clear Water Bay, Kowloon, Hong Kong SAR
dwood@cs.ust.hk.

Abstract. We investigate the complexity of a variety of normal-form transformations for extended context-free grammars, where by extended we mean that the set of right-hand sides for each nonterminal in such a grammar is a regular set. The study is motivated by the implementation project GraMa which will provide a C++ toolkit for the symbolic manipulation of context-free objects just as Grail does for regular objects. The results are that all transformations of interest take time linear in the size of the given grammar giving resulting grammars that are larger by a constant factor than the original grammar. Our results generalize known bounds for context-free grammars but do so in nontrivial ways. Specifically, we introduce a new representation scheme for extended context-free grammars (the symbol-threaded expression forest), a new normal form for these grammars (dot normal form) and new regular expression algorithms.

1 Introduction

In the 1960's, extended context-free grammars were introduced, using Backus–Naur form, as a useful abbreviatory notation that made context-free grammars easier to write. More recently, the Standardized General Markup Language (SGML) [14] used a similar abbreviatory notation to define extended context-free grammars for documents. Currently, XML [5], which is a simplified version of SGML, is being promoted as the markup language for the web, instead of HTML

---

* This research was supported under a grant from the Research Grants Council of Hong Kong SAR. It was carried out while the first and second authors were visiting HKUST.

(a specific grammar or DTD specified using SGML). These developments led to the investigation of how notions applicable to context-free grammars could be carried over to extended context-free grammars. There does not appear to have been any consolidated effort to study extended context-free grammars in their own right. We begin such an investigation with the most basic problems for extended context-free grammars: reduction and normal-form transformations. There has been some related work that is more directly motivated by SGML issues; see the proof of decidability of structural equivalence for extended context-free grammars [3] and the demonstration that SGML exceptions do not add expressive power to extended context-free grammars [15].

We are currently designing a manipulation system toolkit GraMa for extended context-free grammars, pushdown machines, and context-free expressions. It is an extension of Grail [18,17], a similar toolkit for regular expressions and finite-state machines. As a result, we need to choose appropriate representations of grammars and machines that admit efficient transformation algorithms (as well as other algorithms of interest). More specifically we study the extensions specified by regular expressions as used in SGML [14] and XML [5] DTDs.

Earlier results on context-free grammars were obtained by Harrison and Yehudai [10,11,24] and by Hunt *et al.* [13] among others. Harrrison's chapter [10] on normal form transformations provides an excellent survey of early results.

Cohen and Gotlieb [4] suggested a specific representation for context-free grammars and demonstrated how it aided the programming of various operations on them.

We first define extended context-free grammars using the notion of production schemas that are based on regular expressions. In a separate paper [7], we discuss the algorithmic effects of basing the schemas on finite-state machines. As finite-state machines and regular expressions are both first-class objects in Grail, they can be used interchangeably and we expect that they will be in GraMa.

We then describe algorithms for the fundamental normal-form transformations in Section 3. Before doing so, we propose a representation for extended context-free grammars as regular expression forests with symbol threads. We then discuss some algorithmic problems for regular expressions before tackling the various normal forms. When we discuss unit-production removal we will modify this representation somewhat to ensure that we avoid blow-up in the size of the target grammar.

2   Notation and Terminology

We treat extended context-free grammars as context-free grammars in which the right-hand sides of productions are regular expressions. Let V be an alphabet. Then, we define a regular expression over V and its language in the usual way [1,23] with the Kleene plus as an additional operator. The symbol λ denotes the null string. We denote by V_E the set of symbols of V that appear in a regular expression E.

An **extended context-free grammar** G is specified by a tuple (N, σ, P, S), where N and σ are disjoint finite alphabets of **nonterminal symbols** and **terminal symbols,** respectively, P is a finite set of **production schemas,** and the nonterminal S is the **sentence symbol.** Each production schema has the form $A \rightarrow E_A$, where A is a nonterminal and E_A is a regular expression over $V = N \cup \sigma$. When $\beta = \beta_1 A \beta_2 \in V^*$, for some nonterminal A, $A \rightarrow E_A \in P$, and $\alpha \in L(E)$, the string $\beta_1 \alpha \beta_2$ can be derived from the string β and we denote this fact by writing $\beta \Rightarrow \beta_1 \alpha \beta_2$. The **language L(G) of an extended context-free grammar G** is the set of terminal strings derivable from the sentence symbol of G. Formally, $L(G) = \{w \in \sigma^* \mid S \Rightarrow^+ w\}$, where $\Rightarrow^+$ denotes the transitive closure of the derivability relation.

Even though a production schema may correspond to an infinite number of ordinary context-free productions, it is known that extended and standard context-free grammars describe exactly the same languages; for example, see the texts of Salomaa [21] and of Wood [23].

As is standard in the literature, we denote the size of a regular expression E by $|E|$ and define it as the number of symbols and metasymbols in E. We denote the size of a set A by $\#A$. To measure the complexity of any grammatical transformation we need to define the size of a grammar. There are two traditional measures of the size of a context-free grammar that we extend to extended context-free grammars as follows. Given an extended context-free grammar $G = (N, \sigma, P, S)$, we define the **size of G** to be

$$\sum_{A \in N} (1 + |E_A|)$$

and we denote it by $|G|$. We define the **norm of G** to be

$$|G| \log \#(N \cup \sigma)$$

and we denote it by $\| G \|$. Clearly, the norm is a more realistic measure of a grammar's size as it takes into account the size of the encoding of the symbols of the grammar but we use only the size measure here.

3   Normal-Form Transformations

We need the notion of an expression forest that represents the regular expressions that appear as right-hand sides of production schema. Each production schema's right-hand side is represented as an expression tree in the usual way, internal nodes are labeled with operators and external nodes are labeled with symbols. In addition, we represent the nonterminal left-hand side of a production schema with a single node labeled with that nonterminal. The node also refers to the root of the expression tree of its corresponding right-hand side.In Fig. 1, we give an example of forest for two regular expressions.

As an extended context-free grammar has a number of production schemas that are regular expressions, we represent such grammars as an expression forest, where each tree in the forest corresponds to one production schema and each tree is named by its corresponding nonterminal. (The naming avoids the tree repetition problem.) We now add threads such that the thread for symbol X connects all appearances of the symbol X in the expression trees.

3.1   Reachability and Usefulness

A symbol X is **reachable** if it appears in some string derived from the sentence symbol; that is, if there is a derivation $S \Rightarrow^+ \alpha X \beta$ where α and β are possibly null strings over $\sigma \cup N$.

As in standard context-free grammars, reachable symbols can be easily identified by means of a digraph traversal. More precisely, we construct a digraph whose vertices are symbols in $N \cup \sigma$ and there is an edge from A to B if and only if B labels an external node of the expression tree named A. (We assume that the regular expressions do not contain the empty-set symbol.) Then, a depth-first traversal of this digraph starting from S gives all reachable symbols of the grammar. The time taken by the traversal and digraph construction is linear in the size of the grammar.

A nonterminal symbol A is **useful** if there is a derivation $A \Rightarrow^+ \alpha$, where α is a terminal string. The set of useful symbols can be computed recursively as follows. We first find all symbols B such that $L(E_B)$ contains a string of terminal symbols (possibly the null string). All these B are useful symbols. Then, a symbol A is useful if $L(E_A)$ contains a string of terminals and the currently detected reachable symbols, and so on until no newly useful symbols are identified. We formalize this recursive process with a marking algorithm such as described by Wood [22] for context-free grammars. The major difference between previous work and the approach taken here is that we want to obtain an efficient algorithm. Yehudai [24] designed an efficient algorithm for determining usefulness for context-free grammars; our approach can be viewed as a generalization of his work.

To explain the marking algorithm, we assume that we have one bit available at each node of the expression forest to indicate the marking. We initialize these bits in a preorder traversal of the forest as follows: The bits of all nodes are set to zero (unmarked) except for nodes that are labeled with Kleene star, a terminal symbol or the null string—the bits of these nodes are set to one (marked). In the algorithm, whenever a node u is marked (that is, it is useful), it satisfies the condition: *The language of the subtree rooted at u contains a string that is completely marked.* A *-node is marked since its subtree's language contains the null string; therefore, a *-node is always useful.

After completing the initial marking, we propagate markings up the trees in a propagation phase as follows: Repeatedly examine newly marked nodes until no newly marked nodes are obtained. For each newly marked node u, where $p(u)$ is u's parent if u has one, perform one of the following actions:

$p(u)$ **is a +-node** and $p(u)$ is not marked, then mark $p(u)$.
$p(u)$ **is a •-node**, $p(u)$ is not marked and u's sibling is marked, then mark $p(u)$.
$p(u)$ **is a *-node**, then do nothing.

u **is a root node** and the expression tree's nonterminal symbol is not marked, then mark the expression tree's nonterminal symbol.

If there are newly marked nonterminals after this initial round, then we mark all their appearances in the expression forest; otherwise, we terminate the algorithm. We now repeat the propagation phase which propagates the markings of the newly marked symbols up the trees.

The algorithm has, therefore, a number of rounds and at the beginning of each round it marks all appearances of newly discovered useful symbols (discovered in the previous round) and then carries out the propagation phase. As long as a round discovers newly marked nonterminals, the process is repeated. To implement this process efficiently, we construct, at the beginning of each round, a queue of newly marked nodes. Note that the queue is a catenation of appearance lists after the first round. The algorithm then repeatedly deletes an appearance from the queue and, using the preceding propagation rules, it may also append new appearances to the queue. A round terminates when the queue is empty. We can exploit the representation of the grammar by an expression forest to ensure that this algorithm runs in time linear in $|G|$. The idea is to make the test on $L(E_A)$ a pruning operation in the corresponding expression tree. More precisely, each time we test an $L(E_A)$, we delete part of its expression tree; once we "delete" the root, we will have discovered that A is useful. Since the algorithm actually deletes portions of the trees, we apply it to a copy of the expression forest. We first preprocess the forest to prune all subtrees that have Kleene star as their root label.More precisely, we perform a preorder traversal of each tree and whenever we find a *-node we replace it and its subtree with a λ-node. Then, we perform a second pruning in a bottom-up fashion starting from external nodes labeled with the null string, terminal symbols or currently detected useful symbols. This operation is defined recursively as follows. Let x be a node of an expression tree and let $p(x)$ be its parent. If $p(x)$ is a +-node, then delete the subtree rooted at $p(x)$ and recursively apply pruning at $p(p(x))$; if $p(x)$ is a ·-node, then replace it and its subtree with the sibling of x and its subtree, and recursively apply pruning at this node. The overall time for pruning is linear in $|G|$ since we can use the threads to find all appearances of the symbol we want to prune in the expression forest. Note that each node of the expression forest is visited at most twice. Thus, the marking algorithm runs in $O(|G|)$ time and space.

A grammar G is **reduced** if all its symbols are both reachable and useful. As in standard context-free grammars, to reduce a grammar we first identify all useful symbols and then select (together with the corresponding schemas) those that are also reachable.

3.2   Null-Free Form

Since production schemas are not productions but only specify productions, we can convert an extended context-free grammar into a null-free extended context-free grammar much more easily than the corresponding conversion for context-free grammars. Given an extended context-free grammar $G = (N, \sigma, P, S)$, we can determine the nullable nonterminals (the ones that derive the null string) using a similar algorithm to the one we used for usefulness, see Section 3.1. This algorithm takes $O(|G|)$ time. Given this information we can make the grammar null free in two steps. First, we replace all appearances of each nullable symbol A with the regular expression $(A + \lambda)$. This step takes time proportional to the total number of appearances of nullable symbols in G—we use the symbol threads for fast access to them.

In the second step we transform each production schema $A \rightarrow E_A$, where A is nullable, into a null-free production schema $A \rightarrow E'_A$, where $\lambda \notin L(E'_A)$. This step takes time $O(2^{|G|})$ in the worst case. The exponential behavior is attained in a pathological worst case when we have nested dotted subexpressions in which each operand of the dot can produce the null string. We can replace each such pair $F \cdot G$ with $F_- + G_- + (F_- \cdot G_-)$, where F_- is the transformed version of F that does not produce the null string. Note that we double at least the length of the subexpression. As the doubling can occur in the subexpressions of F and G and of their subexpressions, we obtain the exponential worst-case bound. (Note that this is the same case that occurs when the grammar is context-free and every symbol is nullable.)

We can improve the exponential worst-case bound by modifying the transformation of $F \cdot G$ to be $F_- + (F \cdot G_-)$. We can then prove that the size of the resulting expression is, in the worst case, quadratic in the size of the original expression. Indeed, we can apply Ziadi's approach [25] and replace $F \cdot G$ with $F_- + (F \cdot G_-)$, if $|F| \leq |G|$, or with $G_- + (F_- \cdot G)$, otherwise. In this case, we get a blowup of $O(n \log n)$, where n is the size of the original expression. Observe that if we represent the regular expressions with finite-state machines, we would obtain at most quadratic blow up. To reduce the size of the resulting expression to linear, we have to use a different technique invented by We can, however, apply a similar sleight-of-hand technique Yehudai [11,24]. For standard context-free grammars, he reduces the lengths of the right-hand sides to at most two, so that the problem with catenation has only a constant blow up.

We can replace each dotted subtree of an expression tree, that has a dotted ancestor, with a new nonterminal and add a new production schema to the grammar. We can repeat this local modification until no dotted subtree has a dotted subtree within it. Thus, the new grammar has size that is of the same order as the original size. For example, given the production schema

$$A \rightarrow (((a + b + \lambda) \cdot (b + \lambda)) \cdot (c + d + \lambda)) \cdot (e + \lambda);$$

we can replace it with new production schemas:

$$A \rightarrow (B \cdot (c + d + \lambda)) \cdot (e + \lambda)$$

and

$$B \rightarrow (a + b + \lambda) \cdot (b + \lambda).$$

Repeating the transformation for A, we obtain

$$A \rightarrow C \cdot (e + \lambda)$$

and

$$C \to B \cdot (c + d + \lambda).$$

We say that the resulting grammar is in **dot normal form.** We can readily argue that the size of the resulting grammar is at most twice the size of the original size and it has increased the number of nonterminals by at most the size of the original grammar.

Basing the preceding null-removal construction on grammars in dot normal form, ensures that it runs in time $O(|G|)$ time and produces a new grammar that has size $O(|G|)$.

3.3   Unit-Free Form

A **unit production** is a production of the form $A \to B$. Transforming an extended context-free grammar into unit-free form requires three steps. First, we identify all unit-productions. Second, we transform the regular expressions (their representations as expression trees) such that their languages do not include unit nonterminal strings. Lastly, we modify the production schemas such that if $A \Rightarrow^+ B$ (B is unit-reachable from A), then $A \to E'_A + E_B$ where B is not in $L(E'_A)$.

Given an extended context-free grammar G, for each nonterminal symbol A, we can determine its unit productions in time $O(|E_A|)$ by traversing E_A in preorder. Whenever we meet a Kleene star*- or a +-node we continue the traversal in the corresponding subtrees. If we meet a ·-node, then it depends on whether the languages corresponding to the two subtrees contain the null string or not: if one of them does, then we continue the traversal in the other subtree (if both of them contain the null string, then we continue the traversal in both subtrees). When eventually we reach a node labeled with a nonterminal B, then that occurrence of B corresponds to a unit production for A. We define a new thread, which we refer to as the "unit thread" that connects all occurrences of nonterminals that correspond to unit productions in the expression forest. The unit threads are used in the second step. Moreover, while discovering unit productions, we define, for each nonterminal A, the set of nonterminals that are unit-reachable from A. These sets are needed to modify the production schemas. The overall running time for this step is $O(|G|)$.

In the second step, we modify the expression trees at the nodes where we found unit productions. We traverse the expression trees from their frontiers to their roots and, in particular, we follow the paths that start from the nodes labeled with nonterminals that correspond to unit productions (we access them by following the unit threads). Assume, for semplicity that the expression trees do not contain Kleene "*"(if the language of a subexpression contains the null string then we transform the expression using Kleene "+" and symbol λ) and containing λ expression trees Let B_i corresponding to one of these unit production in the expression tree named A. We go up the paths as long we meet only +-nodes.

Consider what happens when we reach a ·-node and its subtrees both yield the null string. Let L and R be its subtrees and let $B = \{B_1, \dots, B_h\}$ and $C = \{C_1, \dots, C_k\}$ be the occurrences of nonterminals corresponding to unit productions that are in L and R, respectively, that we have not yet "eliminated". We replace the ·-subtree with a new tree that combines with +- and ·-nodes the following copies of the original subtrees: L, R, L without B, R without C, L without λ and R without λ. Using the notation $[L], [R], [L - B], [R - C], [L - \lambda]$ and $[R - \lambda]$ to denote these subtrees, we construct the expression:

$$[L - B] + [R - C] + ([L] \cdot [R - \lambda]) + ([L - \lambda] \cdot [R]).$$

In this way we eliminate all unit nonterminal strings $B \cup C$ from the language of the regular expression. When only one of the subtrees yields the null string, the transformation is simpler but similar.

When we meet a *-node, we replace its subtree in a similar way. Let T be a subtree with a *-root node and let $B = \{B_1, \dots, B_h\}$ be the occurrences of nonterminals corresponding to unit productions that we have not yet eliminated. We replace that *-node with a subtree that combines with +- and Kleene $^+$-nodes the following subtrees: λ, T and T without B to give the regular expression

$$\lambda + [T - B] + [T]^{++}.$$

The time complexity of this second step is the same as that of the transformation of regular expressions into null-free expressions (see Section 3.2)

Finally, we need to modify the production schemas such that, for each nonterminal A, if $B_1, \dots, B_h$ are all symbols unit reachable from A, then we define the new production schema as

$$A \rightarrow E_A + E_{B_1} + \dots + E_{B_h}.$$

Rather than duplicating the corresponding trees in the expression forest (it would give a quadratic blow up in the size of the resulting grammar in terms of $|G|$) we define the new expression tree with +-nodes and pointers to the trees named A and $B_1, \dots, B_h$, respectively. This takes time $O(\#N^2)$ and only $O(\#N^2)$ extra space is added to the size of the grammar G.

3.4   Greibach Form

This normal form result for context-free grammars was established by Sheila Greibach [8] in the 1960's; it was a key result in the use of the multiple-path syntactic analyzer developed at Harvard University at that time. An extended context-free grammar is in *Greibach normal form* if its productions have only the following form:

$$A \rightarrow a\alpha,$$

where a is a terminal symbol and α is a possibly empty string of nonterminal symbols. The transformation of an extended context-free grammar into Greibach normal form requires two giant steps: left-recursion removal and back left substitution. Recall that a grammar is left recursive if there is a nonterminal A such that $A \Rightarrow^+ A\alpha$ in the grammar, for some string α. We consider the second step first.

We assume that the given extended context-free grammar $G = (N, \sigma, P, S)$ is reduced, factored, null free, and unit free. In addition, for the second step we also assume that the grammar is not left recursive. Since the grammar is not left recursive there is a partial order on the nonterminals, **left reachability,** that is defined by $A \leq B$ if there is a leftmost derivation $A \Rightarrow^* B\alpha$. As usual, we can consider the nonterminals to be enumerated as $A_1, \ldots, A_n$ such that whenever $A_i \leq A_j$, then $i \leq j$. Observe that A_n is already in Greibach normal form since it has only productions of the form $A_n \to a$, where $a \in \sigma$. We now convert the nonterminals one at a time from A_{n-1} down to A_1. The conversion is conceptually simple, yet computational expensive. When converting A_i, we replace all nonterminals that can appear in the first positions in the strings in $L(E_{A_i})$ with their regular expressions. Thus, $A_i \to E'_{A_i}$ is now in Greibach normal form. To be able to carry out this substitution efficiently we first convert each regular expression E_{A_i} into **first normal form;** that is, we break into the sum of regular expressions that each begin with a unique symbol. More precisely, letting $\sigma = \{a_{n+1}, \ldots, a_{n+m}\}$ and using the notation E_i instead of E_{A_i} for simplicity, we replace E_i by $\bar{E}_i$ which is defined as follows:

$$\bar{E}_i = A_{i+1} \cdot E_{i,i+1} + \cdots + A_n \cdot E_{i,n}$$
$$+ \delta^i_{n+1} \cdot a_{n+1} + \cdots + \delta^i_{n+m} \cdot a_{n+m},$$

where $\sigma = \{a_{n+1}, \ldots, a_{n+m}\}$, $\delta^i_{n+k} = \{\lambda\}$, if $a_{n+k} \in L(E_i)$, and $\delta^i_{n+k} = \emptyset$, if $a_{n+k} \notin L(E_i)$. The conversion should satisfy $L(\bar{E}_i) = L(E_i)$. Then, to convert $\bar{E}_i$ into an equivalent regular expression E'_i in Greibach normal form, we need only replace the first A_k of each term with E'_k.

If the grammar is left recursive, we first need to remove or change left recursion. We use a technique introduced by Greibach [9], investigated in detail by Hotz and his co-workers [12,19,20] and rediscovered by others [6]. It involves producing, for each nonterminal, a distinct subgrammar of G that is left linear and, hence, can be converted into an equivalent right linear grammar. This conversion changes left recursion into right recursion and does not introduce any new left recursion. For more details, see Wood's text [23]. The important property of the left-linear subgrammars is that every sentential leftmost derivation sequence in G can be mimicked by a sequence of leftmost derivation sequences, each of which is a sentential leftmost derivation sequence in one of the left-linear subgrammars. Once we convert the left-linear grammars into right- linear grammars this property is weakened in that we mimic the original derivation sequence with a sequence of sentential rightmost derivation sequences in the right-linear grammars. The new grammar that is equivalent to G is the collection of the distinct right-linear grammars, one for each nonterminal in G.

As the modified grammar is no longer left recursive, we can now apply left back substitution to obtain a final grammar in Greibach normal form.

How well does this algorithm perform? The back left substitution causes, in the worst case, a blow up of $\#N|G|$ in the size of the Greibach normal form grammar. Left recursion removal causes a blow up of $\#N|G|$ at worst. Lastly, converting the production schemas into first normal form causes a blow up of $2\#N|G|$ (this bound is not obvious). As all three operations take time proportional to the sizes of their output grammars, essentially, Greibach normal form takes $O(\#N^5|G|^2)$ time, in the worst case. The reason for the $\#N^5$ term is that we first remove left recursion which not only increases the size of the grammar but also squares the number of nonterminals from $\#N$ to $\#N^2$! The number of nonterminals is crucial in the size bound for the grammar obtained by first normal form conversion and left back substitution.

We can however, reduce the worst-case time and space by using indirection as we did for unit-production removal. Rather than performing the back left substitution for a specific nonterminal, we use a reference to its schema. This technique gives a blowup of only $|G| + \#N^2$ at most and, thus, it reduces the complete conversion time to be $O(\#N^3|G|)$ in the worst case.

We can also apply the technique that Koch and Blum [16] suggested; namely, leave unit-production removal until after we have obtained Greibach-like normal form. Moreover, transforming an extended context-free grammar into dot normal form appears to be a very useful technique to avoid undesirable blow up in grammar size. We are currently investigating this and other approaches.

4   Final Remarks

The results that we have presented are truly a generalization of similar results for context-free grammars. The time and space bounds are similar when relativized to the grammar sizes. The novelty of the algorithms is four-fold. First, we have introduced the regular expression forest with symbol threads as an efficient data representation for context-free grammars and extended context-free grammars. We believe that this representation is new. The only previously documented representations are those of Cohen and Gotlieb [4] and of Barnes [2] and they are much more simplistic. Second, we have demonstrated how indirection using referencing can save time and space in the null-production removal and left back substitution algorithms. Although the use of the technique is novel in this context, it is well known technique in other applications. It is an application of lazy evaluation or evaluation on demand. Third, we have introduced dot normal form for extended context-free grammars that plays a role similar to Chomsky normal form for standard context-free grammars. Fourth, we have generalized the left-linear grammar to right-linear conversion to the extended case.

We are currently investigating whether we can get Greibach normal form more efficiently and whether we can improve the performance of unit- production removal.

Lastly, we would like to mention some other applications of the regular expression forest with symbol threads. First, we can reduce usefulness determination to nullability determination. Given an extended context-free grammar

$G = (N, \sigma, P, S)$, we can replace every appearance of every terminal symbol with the null string to give $G' = (N, \emptyset, P', S)$. Now, a nonterminal A in G is useful if and only if it is nullable in G'. Second, we can use the same algorithm to determine the length of the shortest terminal strings generated by each nonterminal symbol. The idea is that we replace each appearance of a terminal symbol with the integer 1 and each appearance of the null string with 0. We then repeatedly replace: each node labeled "+" that has two integer children with the minimum of the two integers; each node labeled "·" that has two integer children with the sum of the two integers; and each node labeled "*" with 0. The root value is the required length.

We can use the same "generic" algorithm to compute the smallest terminal alphabet for the terminal strings derived from a nonterminal and so on.

References

1. A.V. Aho and J.D. Ullman. *The Theory of Parsing, Translation, and Compiling, Vol. I: Parsing.* Prentice-Hall, Inc., Englewood Cliffs, NJ, 1972.
2. K.R. Barnes. Exploratory steps towards a grammatical manipulation package (GRAMPA). Master's thesis, McMaster University, Hamilton, Ontario, Canada, 1972.
3. H.A. Cameron and D. Wood. Structural equivalence of extended context-free and extended E0L grammars. In preparation, 1998.
4. D.J. Cohen and C.C. Gotlieb. A list structure form of grammars for syntactic analysis. *Computing Surveys*, 2:65–82, 1970.
5. D. Connolly. W3C web page on XML. http://www.w3.org/XML/, 1997.
6. A. Ehrenfeucht and G. Rozenberg. An easy proof of Greibach normal form. *Information and Control*, 63:190–199, 1984.
7. D. Giammarresi and D. Wood. Transition diagram systems and normal form transformations. In *Proceedings of the Sixth Italian Conference on Theoretical Computer Science*, Singapore, 1998. World Scientific Publishing Co. Pte. Ltd.
8. S.A. Greibach. A new normal form theorem for context-free phrase structure grammars. *Journal of the ACM*, 12:42–52, 1965.
9. S.A. Greibach. A simple proof of the standard-form theorem for context-free grammars. Technical report, Harvard University, Cambridge, MA, 1967.
10. M.A. Harrison. *Introduction to Formal Language Theory.* Addison-Wesley, Reading, MA, 1978.
11. M.A. Harrison and A. Yehudai. Eliminating null rules in linear time. *Computer Journal*, 24:156–161, 1981.
12. G. Hotz. Normal-form transformations of context-free grammars. *Acta Cybernetica*, 4:65–84, 1978.
13. H.B. Hunt III, D.J. Rosenkrantz, and T.G. Szymanski. On the equivalence, containment and covering problems for the regular and context-free languages. *Journal of Computer and System Sciences*, 12:222–268, 1976.
14. ISO 8879: Information processing—Text and office systems— Standard Generalized Markup Language (SGML), October 1986. International Organization for Standardization.

15. P. Kilpeläinen and D. Wood. SGML and exceptions. In C. Nicholas and D. Wood, editors, *Proceedings of the Third International Workshop on Principles of Document Processing (PODP 96)*, pages 39–49, Heidelberg, 1997. Springer-Verlag. Lecture Notes in Computer Science 1293.

16. R. Koch and N. Blum. Greibach normal form transformation revisited. In Reischuk and Morvan, editors, *STACS'97 Proceedings*, pages 47–54, New York, NY, 1997. Springer-Verlag. Lecture Notes in Computer Science 1200.

17. D. R. Raymond and D. Wood. Grail: Engineering automata in C++, version 2.5. Technical Report HKUST-CS96-24, Department of Computer Science, Hong Kong University of Science & Technology, Clear Water Bay, Kowloon, Hong Kong, 1996.

18. D.R. Raymond and D. Wood. Grail: A C++ library for automata and expressions. *Journal of Symbolic Computation*, 17:341–350, 1994.

19. R.J. Ross. *Grammar Transformations Based on Regular Decompositions of Context-Free Derivations*. PhD thesis, Department of Computer Science, Washington State University, Pullman, WA, USA, 1978.

20. R.J. Ross, G. Hotz, and D.B. Benson. A general Greibach normal form transformation. Technical Report CS-78-048, Department of Computer Science, Washington State University, Pullman, WA, USA, 1978.

21. A. Salomaa. *Formal Languages*. Academic Press, New York, NY, 1973.

22. D. Wood. *Theory of Computation*. John Wiley & Sons, Inc., New York, NY, 1987.

23. D. Wood. *Theory of Computation*. John Wiley & Sons, Inc., New York, NY, second edition, 1998. In preparation.

24. A. Yehudai. *On the Complexity of Grammar and Language Problems*. PhD thesis, University of California, Berkeley, CA, 1977.

25. D. Ziadi. Regular expression for a language without empty word. *Theoretical Computer Science*, 163:309–315, 1996.

On Parsing LL-Languages

Norbert Blum

Informatik IV
Universität Bonn
Römerstr. 164, D-53117 Bonn, Germany
blum@cs.uni-bonn.de

Abstract. Usually, a parser for an $LL(k)$-grammar G is a deterministic pushdown transducer which produces a leftmost derivation for a
given input string $x \in L(G)$. Ukkonen [5] has given a family of $LL(2)$-
grammars proving that every parser for these grammars has exponential
size. If we add to a parser the possibility to manipulate a constant number of pointers which point to positions within the constructed part of
the leftmost derivation and to change the output in such positions, we
obtain an extended parser for the $LL(k)$-grammar G. Given an arbitrary
$LL(k)$-grammar G, we will show how to construct an extended parser of
polynomial size manipulating at most k^2 pointers.

1 Definitions, Notations, and Basic Results

We assume that the reader is familiar with the elementary theory of $LL(k)$–
parsing as written in standard text books (see e.g. [1,3,6]). First, we will review
the notations used in the subsequence.

A *context-free grammar* (cfg) G is a four-tuple (V, Σ, P, S) where V is a
finite, nonempty set of symbols called the *total vocabulary*, $\Sigma \subset V$ is a finite set
of *terminal symbols*, $N = V \setminus \Sigma$ is the set of *nonterminal symbols* (or *variables*),
P is a finite set of *rules* (or *productions*), and $S \in N$ is the *start symbol*. The
productions are of the form $A \to \alpha$, where $A \in N$ and $\alpha \in V^*$. α is called
alternative of A. $L(G)$ denotes the *context-free language* generated by G. The
size $|G|$ of the cfg G is defined by

$$|G| = \sum_{A \to \alpha \in P} lg(A\alpha),$$

where $lg(A\alpha)$ is the length of the string $A\alpha$. Let ε denote the empty word.

A derivation is *leftmost* if at each step a production is applied to the leftmost
variable. A sentential form within a leftmost derivation is called *left sentential
form*. A context-free grammar G is *ambiguous* if there exists $x \in L(G)$ such
that there are two distinct leftmost derivations of x from the start symbol S. A
context-free grammar $G = (V, \Sigma, P, S)$ is *reduced* if $P = \emptyset$ or, for every $A \in V$,
$S \Rightarrow^* \alpha A\beta \Rightarrow^* w$ for some $\alpha, \beta \in V^*, w \in \Sigma^*$.

J.-M. Champarnaud, D. Maurel, D. Ziadi (Eds.): WIA'98, LNCS 1660, pp. 13–21, 1999

A *pushdown automaton* M is a seven-tupel $M = (Q, \Sigma, \Gamma, \delta, q_0, Z_0, F)$, where Q is a finite, nonempty set of *states*, Σ is a finite, nonempty set of *input symbols*, Γ is a finite, nonempty set of *pushdown symbols*, $q_0 \in Q$ is the *initial state*, $Z_0 \in \Gamma$ is the *start symbol* of the pushdown store, $F \subseteq Q$ is the set of *final states*, and δ is a mapping from $Q \times (\Sigma \cup \{\varepsilon\}) \times \Gamma$ to finite subsets of $Q \times \Gamma^*$.

Given any context-free grammar $G = (V, \Sigma, P, S)$, we will construct a pushdown automaton M_G with $L(M_G) = L(G)$. For the construction of M_G the following notation is useful.

A production in P with a dot on its right side is an *item*. More exactly, let $p = X \to X_1 X_2 \ldots X_{n_p} \in P$. Then (p, i), $0 \leq i \leq n_p$ is an *item* which is represented by $[X \to X_1 X_2 \ldots X_i \cdot X_{i+1} \ldots X_{n_p}]$. Let $H_G = \{(p, i) \mid p \in P, 0 \leq i \leq n_p\}$ be the set of all items of G. Then $M_G = (Q, \Sigma, \Gamma, \delta, q_0, Z_0, F\}$ is defined by

$$Q = H_G \cup \{[S' \to .S], [S' \to S.]\},$$
$$q_0 = \{[S' \to .S]\},\ F = \{[S' \to S.]\},$$
$$\Gamma = Q \cup \{\bot\},\ Z_0 = \bot,\ \text{and}$$
$$\delta : Q \times (\Sigma \cup \{\varepsilon\}) \mapsto 2^{Q \times \Gamma^*}.$$

δ will be defined such that M_G simulates a leftmost derivation. With respect to δ, we distinguish three types of steps.

(E) *expansion*

$$\delta([X \to \beta \cdot A\gamma], \varepsilon, Z) = \{([A \to \cdot\alpha], [X \to \beta \cdot A\gamma]Z) \mid A \to \alpha \in P\}.$$

The leftmost variable in the left sentential form is replaced by one of its alternatives. The pushdown store is expanded.

(C) *reading*

$$\delta([X \to \varphi \cdot a\psi], a, Z) = \{([X \to \varphi a \cdot \psi], Z)\}.$$

The next input symbol is readed.

(R) *reduction*

$$\delta([X \to \alpha \cdot], \varepsilon, [W \to \mu \cdot X\nu] = \{([W \to \mu X \cdot \nu], \varepsilon)\}.$$

The whole α is derived from X. Hence, the dot can be moved beyond X and the corresponding item can be removed from the pushdown store, getting the new state. Therefore, the pushdown store is reduced.

A pushdown automaton is *deterministic* if for each $q \in Q$ and $Z \in \Gamma$ either

i) $\delta(q, a, Z)$ contains at most one element for each $a \in \Sigma$ and $\delta(q, \varepsilon, Z) = \emptyset$ or

ii) $\delta(q, a, Z) = \emptyset$ for all $a \in \Sigma$ and $\delta(q, \varepsilon, Z)$ contains at most one element.

A deterministic pushdown tranducer is a deterministic pushdown automaton with the additional property to produce an output. More formally, a *deterministic pushdown tranducer* is an eight-tuple $(Q, \Sigma, \Gamma, \Delta, \delta, q_0, Z_0, F)$, where all symbols have the same meaning as for a pushdown automaton except that Δ is a finite *output alphabet* and δ is now a mapping $\delta : Q \times (\Sigma \cup \{\varepsilon\}) \times \Gamma \mapsto Q \times \Gamma^* \times \Delta^*$.

For a context-free grammar $G = (V, \Sigma, P, S)$, an integer k, and $\alpha \in V^*$ $FIRST_k(\alpha)$ contains all terminal strings of length $\leq k$ and all prefixes of length k of terminal strings which can be derived from α in G. More formally,

$$FIRST_k(\alpha) = \{x \in \Sigma^* \mid \alpha \Rightarrow^* xy, y \in \Sigma^* \text{ and } |x| = k \text{ or } \alpha \Rightarrow^* x \text{ and } |x| < k\}.$$

A usual way to represent a finite set of strings is the use of tries. Let Σ be a finite alphabet and $|\Sigma| = s$. A *trie* with respect to Σ is a directed tree $T = (V, E)$ where each node $v \in V$ has outdegree $\leq s$. The outgoing edges of a node v are marked by pairwise distinct elements of the alphabet Σ. The node v represents the string $s(v)$ which is obtained by the concatenation of the edge markings on the unique path from the root r of T to v. An efficient algorithm without the use of fixed-point iteration for the computation of all $FIRST_k$-sets can be found in [2].

Let $G = (V, \Sigma, P, S)$ be a reduced, context-free grammar and k be a positive integer. We say that G is $LL(k)$ if G fulfills the following property: If there are two leftmost derivations

1. $S \Rightarrow^*_{lm} wA\alpha \Rightarrow_{lm} w\beta\alpha \Rightarrow^*_{lm} wx$ and
2. $S \Rightarrow^*_{lm} wA\alpha \Rightarrow_{lm} w\gamma\alpha \Rightarrow^*_{lm} wy$

such that $FIRST_k(x) = FIRST_k(y)$, then $\beta = \gamma$.

The following implication of the $LL(k)$ definition is central for the construction of $LL(k)$-parsers. The proof can be found in [1].

Theorem 1. *A cfg $G = (V, \Sigma, P, S)$ is $LL(k)$ if and only if the following condition holds: If $A \to \beta$ and $A \to \gamma$ are distinct productions in P, then $FIRST_k(\beta\alpha) \cap FIRST_k(\gamma\alpha) = \emptyset$ for all $wA\alpha$ such that $S \Rightarrow^* wA\alpha$.*

A *parser* for an $LL(k)$-grammar G is a deterministic pushdown tranducer which produces a leftmost derivation for a given input $x \in L(G)$. If we add to a parser the possibility to manipulate a constant number of pointers which point to positions within the constructed part of the leftmost derivation and to change the output in such positions, we obtain an *extended parser* for the $LL(k)$-grammar G.

Ukkonen [5] has given a family of $LL(2)$-grammars and shown that every parser for these grammars must have exponential size. Given an arbitrary $LL(k)$-grammar G, we will show how to construct an extended parser of polynomial size manipulating at most k^2 pointers.

2   The Construction of Polynomial Size Extended $LL(k)$-Parser

2.1   A Motivating Example

Given any $LL(k)$-grammar G, our goal is the construction of an extended $LL(k)$-parser of polynomial size for G. In order to explain the main idea, we consider the

$LL(2)$-grammar, given by Ukkonen [5] for proving an exponential lower bound on the size of any parser. For $n \in \mathbb{N}$ consider the cfg $G_n = (V_n, \Sigma_n, P_n, A_0)$ defined by

$$P_n : A_i \to a_{i+1}A_{i+1}B_{i+1} \mid d_{i+1}A_{i+1}C_{i+1} \quad (0 \leq i \leq n-1)$$
$$A_n \to b_i \mid \varepsilon \qquad\qquad (1 \leq i \leq n)$$
$$B_i \to b_i c_i \mid \varepsilon \qquad\qquad (1 \leq i \leq n)$$
$$C_i \to c_i \mid \varepsilon \qquad\qquad (1 \leq i \leq n)$$

It is easy to see that G_n is an $LL(2)$-grammar. Let us first review Ukkonen's idea for proving the exponential lower bound for the size of any $LL(k)$-parser, $k \geq 2$ for G_n. Consider the leftmost derivation of a string $x_1 x_2 \ldots x_n b_j c_j \ldots$, where $x_i \in \{a_i, d_i\}$, $1 \leq i \leq n$. The critical point is when the parser has to determine the correct alternative for A_n with respect to the left sentential form $x_1 x_2 \ldots x_n A_n X_n X_{n-1} \ldots X_1$. It needs the knowledge of whether $x_j = a_j$ or $x_j = d_j$. In the case that $x_j = a_j$, ε would be the correct alternative, since $b_j c_j$ would be derived from B_j. In the other case, c_j would be derived from C_j and hence, b_j would be the correct alternative for A_n. In the case that the parser does not have the possibility to continue the parsing process and to determine the correct alternative for A_n later, there are two possibilities to obtain this information:

1. The information is contained in the actual state q of the parser and the topmost stack symbol Z. But then, the number of distinct pairs (q, Z) must be exponential in n since there exist 2^n distinct strings $x_1 x_2 \ldots x_n$ with $x_i \in \{a_i, d_i\}, 1 \leq i \leq n$, and j is not known in advance. This would imply that the size of the parser must be exponential in the size of the grammar.
2. The parser looks into the stack whether B_j or C_j is pushed. But then, the information above B_j and C_j, respectively must be stored in the current state q and the current topmost stack symbol Z. But then, the number of pairs (q, Z) must be exponential in the size of the grammar, since for the left sentential form $x_1 x_2 \ldots x_n A_n X_n X_{n-1} \ldots X_{j+1} X_j \ldots$ with $X_l \in \{B_l, C_l\}, n \geq l > j$, the number of possible distinct strings $X_n X_{n-1} \ldots X_{j+1}$ is 2^{n-j}. This would also imply that the size of the parser must be exponential in the size of the grammar.

If we allow to continue the parsing process and to delay the determination of the correct alternative for A_n to the future, then we will obtain an extended $LL(2)$-parser for G_n of size $O(n)$. We only need a pointer to the correct position of the production with left hand side A_n in the computed leftmost derivation. The extended parser works in the following way.

- The correct alternatives for the variables $A_0, A_1, \ldots, A_{n-1}$ can be derived immediately from the first unread symbol of the input. The corresponding productions can be written directley into the leftmost derivation under construction.
- Let us consider the moment, when the parser wants to expand A_n. If the first unread symbol of the input is in $\{c_1, c_2, \ldots, c_n\}$ then it is clear that ε

is the correct alternative of A_n. If the first unread symbol of the input is b_j, $1 \leq j \leq n$, and the second unread symbol is $\neq c_j$, then b_j is the correct alternative of A_n. In both cases, the correct production can be written into the leftmost derivation under construction.

If the first and second unread symbols are b_j and c_j, $1 \leq j \leq n$, then the parser cannot determine the correct alternative of A_n. The parser creates a pointer, pointing to the correct (i.e. current) position for the production with left hand side A_n in the leftmost derivation under construction and continues with the construction of the leftmost derivation.

For each variable of $X_n X_{n-1} \ldots X_{j+1}$ it is clear that ε is the correct alternative and the parser writes successively the corresponding production into the leftmost derivation under construction.

Let us consider the moment, when X_j should be expanded. If $X_j = B_j$ then $b_j c_j$ is the correct alternative for X_j and ε is the correct alternative for A_n. Otherwise, i.e., $X_j = C_j$, c_j is the correct alternative of X_j and b_j is the correct alternative of A_n. In both cases, the parser can write the correct production with respect to A_n at the correct position of leftmost derivation under construction. Furthermore, the production corresponding to X_j can be written into the leftmost derivation.

- For all further variables, the correct alternatives can be derived from the first unread symbol of the input. Hence, the corresponding productions can be written directly into the leftmost derivation.

2.2 The General Construction

Our goal is now to generalize the method explained with help of the example above such that it can be used for any $LL(k)$-grammar.

Let $G = (V, \Sigma, P, S)$ be an arbitrary $LL(k)$-grammar. Consider the pushdown automaton M_G for G. Our goal is to construct the extended parser P_G by a step-by-step extension of M_G. For doing this, assume that w is the read input, x is the substring of length k which follows w and that $[X \rightarrow \beta.A\gamma]$, $A \in N$ is the state of the pushdown automaton. Then, P_G wants to expand the variable A. Furthermore, assume that P_G knows no symbol of $x = x_1 x_2 \ldots x_k$.

Our goal is now to determine the prefixes of x which can be derived exactly from A. These can be only prefixes x' of x such that $x' \in FIRST_k(A)$. Hence it is useful that

- P_G computes the maximal prefix u of x which is also prefix of an element of $FIRST_k(A)$.

For an efficient solution of this subgoal, we add for each variable $X \in N$ the trie $T_k(X)$, corresponding to $FIRST_k(X)$ to P_G. Now,

- P_G starts to read the lookahead x and, simultaneously, follows the corresponding path in $T_k(A)$, starting at the root, until the maximal prefix u of x in $T_k(A)$ is determined.

Depending on whether $|u| < |x|$ or $|u| = |x|$, we distinguish two cases.

Case 1 : $|u| < |x|$

It follows directly that the terminal string which is derived form A must be a prefix u' from u. Furthermore, $u' \in FIRST_k(A)$. Hence, it is useful if P_G has direct access to such prefixes. For getting this, every node $v \in T_k(A)$ contains a pointer to each node $w \in T_k(A)$ such that

1. w is on the path from the root of $T_k(A)$ to v, and
2. $s(w) \in FIRST_k(A)$.

Since P_G does not know the whole lookahead x, possibly P_G cannot decide which of these $\leq k$ prefixes is the correct string which is derived from A within the leftmost derivation of the input. Hence, P_G has to store all these possibilities.

For doing this, P_G initializes for each of these prefixes u' a queue which contains one element. This element contains the variable A, the terminal string u', and a pointer to the position within the leftmost derivation at which the derivation of u' from A has to be written if u' would be the correct terminal string.

Now, the parser P_G continues the construction of the leftmost derivation with the first not expanded variable or not read terminal symbol of the state or an item in the pushdown store. For simplicity, we assume that the current state is always stored at the top of the pushdown store. Since for all items, the treatment of the symbols before the dot is finished and the symbol directly behind the dot is a variable which has just been expanded, the needed symbol is at the second position behind the dot of an item. Hence, as long as the topmost item in the pushdown store contains less than two symbols behind the dot, P_G performs pop-operations.

Assume that B is the second symbol behind the dot of the topmost item in the pushdown store. Then B is a nonexpanded variable or B is a terminal symbol which is not treated by a reading step. In both cases, P_G considers iteratively all queues. Assume that M is the queue under consideration and that j is the length of the prefix u of x which corresponds to the queue M. We distinguish two cases.

Case 1.1: $B \in N$

It is useful if P_G knows the maximal prefix of the next $k - j$ input symbols which is also a prefix of an element in $FIRST_k(B)$. In contrast to the above, P_G already read $|u| - j$ of these $k - j$ symbols. Hence, P_G needs the possibility of direct access to the "correct" node in $T_k(B)$ with respect to the read prefix of the next $k - j$ symbols. Then, P_G can process analogously to the above.

For getting this direct access, we extend P_G by a trie T_G representing the set Σ^k. Also, P_G manipulates a pointer $P(T_G)$ which always points to the node u in T_G with $s(u)$ is the prefix of the lookahead x already read.

For $v \in T_G$ let $d(v)$ denote the depth of v in T_G and $s_i(v)$, $0 \leq i < d(v)$ denote the suffix of $s(v)$ which starts with the $(i+1)$st symbol of $s(v)$. Every node $v \in T_G$ contains for all $A \in N$ and $1 \leq i < d(v)$ a pointer $P_{i,A}(v)$ which points to the node $w \in T_k(A)$ such that $s(w)$ is the maximal prefix of an element of $FIRST_k(A)$ which is also a prefix of $s_i(v)$.

Using the pointer $P_{j,B}(u)$, where u is the node to which $P(T_G)$ points, P_G has direct access to the correct node w in $T_k(B)$.

If $s(w) \neq s_j(u)$ then $s(w)$ is the maximal prefix of $s_j(u)$ which is prefix of an element of $FIRST_k(B)$. In this case, P_G proceeds analogously to the above. But instead of the initialization of new queues, P_G extends the queue M under consideration in an appropriate manner. This can be done as follows.

For every prefix u' of $s(w)$ with $u' \in FIRST_k(B)$, the node corresponding to the head of the queue A obtains a new successor corresponding to the prefix u'. If no such u' exists, then P_G is in a dead end with respect to the queue M and M can be deleted completely.

If $s(w) = s_j(u)$ then P_G continues to read the rest of the lookahead x and, simultaneously, follows the corresponding path in $T_k(B)$, starting at the node w.

Case 1.2: $B \in \Sigma$

If $B = x_{j+1}$ then P_G extends the queue M under consideration by adding a new successor corresponding to x_{j+1} to the node corresponding to the head of M. Otherwise, P_G is in a dead end with respect to the queue M and M can be deleted completely.

Case 2: $|u| = |x|$

Then $x \in FIRST_k(A)$. Theorem 1 implies that for every proper prefix u' of x such that $u' \in FIRST_k(A)$ it holds that with respect to the leftmost derivation for the input string of P_G, u' cannot be the terminal string which is derived from A. Moreover, at most one alternative α of A with $x \in FIRST_k(\alpha)$ exists. Hence, P_G can determine the correct expansion step.

Altogether, the data structure for the set of queues is a forest. The path from the root of a tree to a leaf always corresponds to an actual queue and vice versa. The leaf is the head of the queue and the root is the tail of the queue. Every node on the path from the tail to the head corresponds to a substring of x. The prefix of x corresponding to the queue can be obtained by the concatenation of these substrings. As observed below, the corresponding prefixes of two distinct queues are distinct. Moreover, the roots of two distinct trees of the forest correspond to distinct prefixes of x.

Next, we want to derive an upper bound for the number of queues. Note that an $LL(k)$-grammar is always unambiguous.

Lemma 1. *The number of queues is always $\leq k$.*

Proof. Consider the moment, when for the current symbol all work with respect to all the queues is done. Each such a moment corresponds to a string $\gamma = A\gamma' \in NV^*$, where P_G does not know the correct alternative for A. The queues correspond to pairwise distinct derivations of a proper prefix of x from γ. Note that ε is a prefix of length 0 of x.

Assume that two distinct queues correspond to a derivation of the same prefix of x. Then there exists a word in $L(G)$ which has two distinct leftmost derivations from the start symbol S. Note that G is reduced. But this would be a contradiction to the unambiguity of G. Hence, at most one queue corresponds to every prefix of x. Hence, the number of queues is bounded by k.   □

Now we want to bound the length of every queue by k. If we take care that every node up to the head of a queue corresponds to a string of length > 0, then every queue contains at most k nodes. For getting this property, every time when ε can be the string derived from the variable C under consideration, P_G writes the unique derivation of ε from C into the leftmost derivation under construction. In the case that another string u would be derived from C, this wrong derivation will be superscribed later by the correct derivation of u from C. Hence, always when ε is the correct terminal string derived from a variable C, the correct derivation of ε from C is written at the leftmost derivation under construction and nothing has to be changed with respect to this part of the derivation. We need the possibility that the head of a queue can correspond to the terminal string ε for the case, that the father of this head has another son and also for the case that there is a queue which corresponds to the string ε. By construction, nodes corresponding to ε need no pointer to a position within the leftmost derivation under construction. In the case that the head of the queue M under consideration corresponds to ε, instead of the head of M, the father of the head of M obtain new successors.

Altogether, the number of nodes within the forest is bounded by k^2 and hence, the number of pointers is bounded also by k^2.

The earliest moment when the prefix of x derived from A can be determined is the moment when the forest consists of exactly one tree. This prefix corresponds to the root of this unique tree. By Theorem 1, this must happen when the last symbol x_k of the lookahead is read or earlier. When the forest consists of exactly one tree, the prefix u' of x corresponding to the root r of this tree is the terminal string which is derived from A. P_G modifys the leftmost derivation under construction as follows.

- Using the pointer of the root r, P_G has access to the position within the leftmost derivation under construction, where the leftmost derivation of u' from A has to be written. We can assume that this derivation is precomputed such that P_G can write this derivation at the correct position.
- The root r is deleted. Eventually, the new forest contains now more than one tree.

Since u' is derived from A, the new lookahead is shifted $|u'|$ positions. Hence, P_G has to update the pointer $P(T_G)$. For doing this within constant time, each node $v \in T_G$ contains $d(v) - 1$ pointers $P_i(v)$, $1 \le i < d(v)$ pointing to the unique node $w \in T_G$ with $s(w) = s_i(v)$. Now, P_G performs the following steps.

– Using the pointer $P_{|u'|}(v)$, where v is the node to which $P(T_G)$ points, P_G has direct access to the node, to which the pointer $P(T_G)$ has to point. After updating the pointer $P(T_G)$, the parser continues its work at that point, where the work was interrupted.

Next, we want to bound the size of P_G and the parsing time. By construction, P_G contains $|N| + 1$ tries. Each trie consists of at most $2|\Sigma|^k$ nodes. For all nodes in T_G, the number of pointers is bounded by $(|N| + 1)k$. For all $A \in N$ for all nodes in $T_k(A)$, the number of pointers is at most k. Hence, all tries need $O(k|N||\Sigma|^k)$ space. The space for the precomputed leftmost derivations for terminal strings of length $\leq k$ is bounded by $O(|N||\Sigma|^k)$. Note that G is an $LL(k)$-grammar.

For each reading and expansion, respectively at most k queues are considered. Since the number of queues does not exceed k, the whole time per step for the construction of queues is $O(k)$. For the decomposition of queues no more time than for the construction is needed. Hence, the parsing time is bounded by $O(k \cdot$ length of the derivation$)$. Altogether, we have proven the following theorem.

Theorem 2. *Let $G = (V, \Sigma, P, S)$ be an $LL(k)$-grammar. Then there is an extended parser P_G for G which has the following properties.*

i) The size of the parser is $O(k|N||\Sigma|^k))$.
ii) P_G needs only the additional space for at most k^2 pointers.
iii) The parsing time ist bounded by $O(k \cdot$ length of the derivation$)$.

Acknowledgment: I thank Claus Rick for critical remarks.

References

1. A. V. Aho, and J. D. Ullman, *The Theory of Parsing, Translation, and Compiling*, Vol. I: Parsing, Prentice-Hall (1972).
2. N. Blum, *Theoretische Informatik: Eine anwendungsorientierte Einführung*, R. Oldenbourg Verlag München Wien (1998).
3. S. Sippu, and E. Soisalon-Soininen, *Parsing Theory, Vol. I: Languages and Parsing*, EATCS Monographs on Theoretical Computer Science Vol. 15, Springer (1988).
4. S. Sippu, and E. Soisalon-Soininen, *Parsing Theory, Vol. II: LR(k) and LL(k) Parsing*, EATCS Monographs on Theoretical Computer Science Vol. 20, Springer (1990).
5. E. Ukkonen, Lower bound on the size of deterministic parsers, *JCSS* **26** (1983), 153–170.
6. R. Wilhelm, and D. Maurer, *Compiler Design*, Addison-Wesley (1995).

On Parsing and Condensing Substrings of LR Languages in Linear Time*

Heiko Goeman

University of Bonn
Computer Science Department V
Römerstraße 164, 53117 Bonn, Germany
goeman@cs.uni-bonn.de

Abstract. LR parsers have long been known as being an efficient algorithm for recognizing deterministic context–free grammars. In this article, we present a linear–time method for parsing substrings of LR languages. The algorithm depends on the LR automaton that is used for the usual parsing of complete sentences. We prove the correctness and linear complexity of our algorithm and present an interesting extension of our substring parser that allows to condense the input string, which increases the speed when reparsing that string for a second time.

1 Introduction

The problem of recognizing substrings of context–free languages has emerged in several interesting applications and can be described as follows. Given a string y and a grammar $G = (V, \Sigma, P, S)$, we wish to know whether there exist two additional strings x and z such that xyz is a sentence of G. An important application for a corresponding *substring parser* is a method for detecting syntax errors suggested by Richter [5], although his article does not contain such a parser. The ability to decide whether a part of a given program is not a substring of a programming language allows the local detection of syntax errors without performing a complete parsing process.

Several substring parsers suffering from various drawbacks have already been presented before. Cormack's algorithm [4] and a parallel version of it [3] only work with the *bounded–context* class of grammars, which is a proper subset of the LR(1) class. Bates and Lavie's approach [2] is applicable for SLR(1), LALR(1) and all canonical LR(k) grammars, but their correctness proof as well as their complexity analysis are incorrect, as we shall show later.

In this article, we develop a substring parser that can be used with SLR(k), LALR(k) and canonical LR(k) grammars. Even more, our parser determines the maximum prefix of an input string y that represents a substring. We also present another interesting feature called *condensation* of substrings. This feature allows

---

* This work is dedicated to my mother.

J. M. Champarnaud, D. Maurel, D. Ziadi (Eds.): WIA'98, LNCS 1660, pp. 32–42, 1999.

to transform an input string y into a string $\beta \in V^+$ such that the following two conditions are satisfied:

$$\beta \Longrightarrow_G^* y \quad \text{and} \quad \forall x, z \colon xyz \in L(G) \Longrightarrow x\beta z \text{ is a sentential form of } G \ .$$

Thus, all reductions stored in a condensation string β of an input string y must always be done when parsing a string that contains y as a substring. This allows the replacement of y with β and increases the processing speed when reparsing the resulting string because the mentioned reductions are automatically skipped.

We begin in Sect. 2 with a review of the basic terminology and definitions used throughout this paper. The algorithm is then explained in Sect. 3. In this first version, the parser is only applicable to SLR(1), LALR(1) and LR(1) grammars. The correctness of this algorithm is proven in Sect. 4, and Sect. 5 deals with the analysis of its linear time complexity. Section 6 describes the condensation feature, and Sect. 7 deals with the necessary modifications in order to use the substring parser with canonical LR(k), SLR(k) and LALR(k) grammars, where $k > 1$. An appendix pinpoints the already mentioned flaws in Bates and Lavie's work.

2   Terminology and Definitions

In this section, the basic terminology and definitions used in this article are introduced. We assume that the reader is familiar with the LR parsing technique. For more information, the reader is directed to [1].

A *context–free grammar* is a quadruple $G = (V, \Sigma, P, S)$, where V is the set of *grammar symbols* called the *vocabulary*, $\Sigma \subset V$ is a set of *terminal symbols*, $N := V \setminus \Sigma$ is the set of *variables*, P is the set of *productions* (or *rules*), and $S \in N$ is the *start symbol*. A production is of the form $A \to \alpha$, where $A \in N$ and $\alpha \in V^*$. We use $A \to \gamma_1 \mid \ldots \mid \gamma_n$ to denote the productions $A \to \gamma_1, \ldots, A \to \gamma_n$. We assume that G is always unambiguous and reduced, i.e., G does not contain any unnecessary symbols.

Letters used in formulas have the following meaning. Upper and lower case letters at the beginning of the alphabet denote variables and terminals, respectively, whereas upper and lower case letters at the end of the alphabet are general grammar symbols in V and terminal strings in Σ^*, respectively. Greek lower case letters denote vocabulary strings in V^*.

A *substring* of G is a string y such that there exist some x, z with $xyz \in L(G)$, where $L(G)$ denotes the language generated by G. We use $SS(G)$ and $S(G)$ to denote the set of all substrings and the set of all sentential forms, respectively.

Let $k \in \mathbb{N}_0$. A quadruple (A, α, β, x), written $[A \to \alpha_\bullet \beta, \{x\}]$, is called an *LR($k$)–item* of G if $A \to \alpha\beta \in P$ and $x \in \Sigma^{\leq k}$, i.e., $x \in \Sigma^*$ and $|x| \leq k$. An SLR(k), LALR(k) or LR(k) parser always depends on several sets containing these items because these sets make up the states of its corresponding *LR–DFA*, which is a deterministic finite automaton that is used to build up the usual functions *Action* and *Goto*. For example, the appropriate LR–DFA for an SLR(1) grammar consisting of the rules

$$S \longrightarrow Ab \mid aa \mid aAb \qquad A \longrightarrow aa$$

is given in Fig. 1. Its start state, generally denoted by q_0, is positioned in the upper left corner. The emphasized states q_2, q_6 and q_7 are the final states. They all contain an item of the form $S \to \alpha_{\bullet}, \{\varepsilon\}$ which indicates that the parsing process is complete.

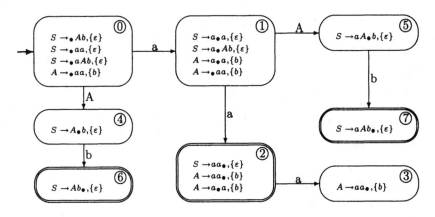

Fig. 1. The state transition diagram of the sample SLR(1) grammar

We denote the set of all states with Q, and the transition function (which in fact represents the function *Goto*) with δ. We extend the domain of δ on sets $S \subseteq Q$ and strings in Γ^* in the usual way by establishing

$$\delta(S, a) := \{\delta(q, a) \mid q \in S\}, \quad \delta(S, \varepsilon) = S \text{ and } \delta(S, aw) = \delta(\delta(S, a), w) \ .$$

All incoming edges of a state q in an LR–DFA are labelled with an unique symbol which we denote by $\varphi(q)$. The corresponding function $\varphi : Q \to V$ can be easily extented to a function $\tilde{\varphi}$ that accepts strings $s_1 \ldots s_r$ in Q^+ by establishing $\tilde{\varphi}(s_1 \ldots s_r) := \varphi(s_2) \ldots \varphi(s_r)$. Note that since s_1 is often equal to q_0 in the following applications of $\tilde{\varphi}$, the first state s_1 is omitted in the definition of this function because q_0 does not have any incoming edges.

A *configuration* of an LR parser describes the status the parser is currently in. More precisely, a configuration is a pair $(s_1 \ldots s_r, a_i \ldots a_n)$, where $s_1, \ldots, s_r$ are the states currently pushed on the stack (with s_r on the stack top), and $a_i \ldots a_n$ is the rest of the input string $a_1 \ldots a_n$ that has not yet been read. Clearly, from the way an LR parser works, s_1 must be equal to q_0, and, for $j = 1, \ldots, r-1$, $\delta(s_j, \varphi(s_{j+1})) = s_{j+1}$.

Only a part of a configuration is used to determine the next parsing step because an LR(k) parser only looks at the next k symbols $a_i, \ldots, a_{\min\{i+k-1,n\}}$. Furthermore, the function *Action* only additionally needs the top state s_r as a parameter. Thus, we can define a *partial configuration* by dropping the condition $s_1 = q_0$ and only demanding that the string part contains at least the next k symbols. The stack part of such a configuration is then called a *partial stack*.

Let k_1, k_2 be two (partial) configurations. We write $k_1 \vdash^S k_2$ and $k_1 \vdash^{A \to \alpha} k_2$ to denote that k_2 results from k_1 due to shifting the next symbol and due to performing a reduction by the rule $A \to \alpha$, respectively. Moreover, $k_1 \vdash^R k_2$ and $k_1 \vdash k_2$ mean that k_2 results from k_1 due to a reduction by some uninteresting rule and due to any single parsing step, respectively. For example, the parsing process of the string $aaab \in L(G)$, where G is the grammer in the example given above, can be described as follows:

$$(q_0, aaab) \vdash^S (q_0 q_1, aab) \vdash^S (q_0 q_1 q_2, ab)$$
$$\vdash^S (q_0 q_1 q_2 q_3, b) \vdash^{A \to aa} (q_0 q_1 q_5, b) \vdash^S (q_0 q_1 q_5 q_7, \varepsilon) .$$

The reflexive and transitive closure of $\vdash$ is denoted by $\vdash^*$. Note that from the way an LR parser works, if $(q_0, vw) \vdash^* (s, w)$, then $S \xRightarrow{\text{rm}}^* \tilde{\varphi}(s)w \xRightarrow{\text{rm}}^* vw$, where $\xRightarrow{\text{rm}}$ denotes the usual rightmost derivation.

3   The Algorithm

We now outline the idea which supports our substring parser. In this first version, it is suitable for any LR–parser with an one–symbol–lookahead (for example, an SLR(1) parser).

Let us assume that y is the input string used with our algorithm. The substring parser simulates the behaviour of the LR parser when processing the part y of some input string xyw. The difference to a regular parsing process is that in this case the configuration the LR parser is in after processing x is now unknown. Moreover, there are usually many different possibilities for x and the corresponding configurations. Our algorithm gets around this problem by managing several partial configurations at the same time which on one hand correspond to these complete configurations, and on the other hand contain all the information that result from parsing the substring y. Using this idea, we now analyse how the partial configurations must look like at the beginning.

After processing the prefix x of the complete input string xyw, the original LR parser starts the parsing process of $y = az$ by shifting a. Clearly, just before this step, the top state q on the stack must satisfy the condition $Action(q, a) = Shift$. Conversely, every state q with this property can in fact make up the topmost state because there always exists some string x and a path p from q_0 to q in the LR–DFA such that $\tilde{\varphi}(p) \xRightarrow{\text{rm}}^* x$ and $(q_0, xaz) \vdash^* (p, az) \vdash^S (pq', z)$, where $q' := \delta(q, a)$. For example, in Fig. 1, let $q := q_4$, and let b be the shifted symbol. Then we can choose $p := q_0 q_4$ and $x := aa$ because x can be derived from $\varphi(p) = A$. As desired, we obtain:

$$(q_0, aabz) \vdash^S (q_0 q_1, abz) \vdash^S (q_0 q_1 q_2, bz) \vdash^{A \to aa} (q_0 q_4, bz) \vdash^S (q_0 q_4 q_6, z) .$$

Thus, our substring parser starts with all partial configurations of the form (q, az), where q satisfies the above condition. Then, the algorithm alternately simultates the shift and reduce operations of the LR parser. Clearly, at first a shift operation must be simulated. Corresponding to the action an LR parser performs in such a situation, this is simply done by pushing a new state $\delta(q,a)$ on every partial stack, where q denotes the respective topmost state and a denotes the next input symbol. The input pointer is then advanced to the next symbol. Thus, during the first shift simulation, every partial configuration (q, az) is replaced with $(q\delta(q,a), z)$. Later shift operations are handled in the same way.

Between two shift operations, the substring parser simulates the corresponding reductions. Provided that the partial stack of a configuration is large enough, this again can be done in the usual way. For example, using Fig. 1, consider the configuration $(q_0q_1q_2, b)$. The LR parser then reduces the stack due to the rule $A \rightarrow aa$, and so does our algorithm. The resulting configuration afterwards is (q_0q_4, b). But when the partial stack does not contain enough states, the algorithm has to take care about all possible extensions of it. For example, let (q_1q_2, b) be the starting configuration. Now the reduction $A \rightarrow aa$ cannot directly be handled because at least one state q is missing at the stack bottom. Clearly, q_1 must be accessible from q because the partial stack always corresponds to a path in the LR–DFA. Thus, q must be equal to q_0, and again (q_0q_4, b) is the resulting configuration. If q_1 had been accessible from another state q', then the substring parser would have generated another configuration $(q'\delta(q', A), b)$. In general, when there are r states missing in the stack and q is the bottom state, the algorithm has to consider the states in $\delta^{-1}(q, r)$, where the function $\delta^{-1} : Q \times \mathbb{N}_0 \rightarrow 2^Q$ is defined as follows:

$$\delta^{-1}(q, r) := \{q' \mid \exists \alpha \in V^r : \delta(q', \alpha) = q\} .$$

Then for every $q \in \delta^{-1}(q, r)$, a new partial configuration with the stack contents $q\delta(q, A)$ is generated.

We now discuss the management of the partial configurations. The substring parser maintains a directed labelled graph $Gr = (V, E, l)$, where V, E and l denote a set of *vertices* (or *nodes*), a set of *edges*, and a *labelling function* $l : V \rightarrow Q$, respectively. The graph structure consists of several trees, and the root nodes of these trees are collected in a set T. We are exclusively interested in the maximum paths contained in the trees (i.e., paths from leafs to root nodes), and therefore from now on, when speaking of a path, we always mean a maximum one. Let p be such a path in Gr and let $|p|$ denote the length of p, i.e., the number of edges in it. Let s_i be the label of the i–th node, $1 \leq i \leq |p| + 1$. Finally, let $y = y_1y_2$, where y_1 is the prefix read so far. Then p represents all configurations (s, y_2z) with the following two properties. Firstly, (s, y_2z) can be obtained by parsing the prefix xy_1 of a sentence xy_1y_2z with the original LR algorithm. Secondly, the *labelling of* p, defined as $l(p) := s_1 \ldots s_{|p|+1}$, is a suffix of s.

Conversely, every configuration with the first property is represented by some path in the graph. Hence, y_1 is a substring of $L(G)$ iff the graph is not empty. Furthermore, a path p can be discarded from the graph if there exists another path p' such that $l(p')$ is a suffix of $l(p)$.

Two alternately called procedures maintain the graph. The first one, *Shift-CommonSymbol (SCS)*, changes the paths in the same manner as the LR algorithm changes the stack portion of a configuration due to a shift operation, i.e., a path that ends in a root node labelled with some state q will be extended by a new node labelled with the state $\delta(q, a)$, where a denotes the shifted symbol. The other procedure, *ReduceStacks (RS)*, simulates reduce operations in a similar way, i.e., for a reduction according to some grammar rule $A \rightarrow \alpha$, at first the substring parser drops $|\alpha|$ nodes and edges from the end of a path and then appends a new node labelled with $\delta(l(v), A)$, where v is the last node of the path after the first step. When a path p contains fewer than $|\alpha|$ edges, it is necessary to replace p by several new *short paths*. Each of them consists of two nodes, and the label q of the first node is a state in $\delta^{-1}(l(v), |\alpha| - |p|)$, where v denotes the first node of p. As in the previous case, the state of the ending node is $\delta(q, A)$. To avoid redundant work, each of these short paths can be produced only once during one execution of RS.

The case *Action = Error* is simulated by simply deleting the corresponding path. The algorithm terminates when either the complete input string has been read or every path has been deleted. In the latter case, the part of the input string read so far is the longest prefix representing a substring.

Figure 2 shows the development of the graph when parsing the substring ab. In this simple example, each tree always consists of only one single path. The vertices with index T are the root nodes.

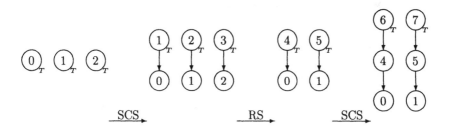

Fig. 2. The graph structure while parsing the substring ab

In order to obtain a linear–time complexity it is essential for the root nodes of the trees to be labelled differently, and this is necessary for the children of any node as well. Assume there are two trees with this property whose root nodes v_1 and v_2 are labelled with the same state. As mentioned above, a certain set of configurations is represented by these trees. The following procedure then merges them into one tree such that the resulting tree has v_1 as its root node and represents the same set of configurations. The procedure uses the fact mentioned earlier that a path p can be removed from the graph if there exists another path such that its labelling is a suffix of $l(p)$.

Procedure MergeTrees(v_1, v_2) {
    If v_1 and v_2 both have children **Then** {
        Disconnect all children from v_2 and delete v_2;
        For each former child c_2 of v_2 **Do** {
            If v_1 has a child c_1 such that $l(c_1) = l(c_2)$ **Then**
                MergeTrees(c_1, c_2)
            Else Let c_2 become a child of v_1;
        }
    } **Else** Remove both trees completely except for the single node v_1;
}

As an example, let us assume we have an LR–DFA with at least nine states, and the two trees on the left side in Fig. 3 have been generated at some time. The result from merging them is then shown on the right side.

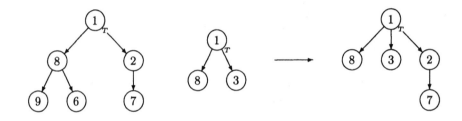

Fig. 3. An example for merging two trees

The given facts lead to the following algorithm. It determines the index j of the maximum prefix $a_1 \ldots a_j \in SS(G)$ of the input string $a_1 \ldots a_n \in \Sigma^+$. Some program lines are marked with a bar ($|$) on their left sides. These lines correspond to the extension that condenses the input string and are explained later in Section 6. For the moment, we simply assume that these lines are not present.

```
1     Procedure ShiftCommonSymbol (SCS) {
          M := ∅;
|         If |T| = 1 Then h := h + 1;
          For v ∈ T Do { (* Extend all paths ending in v *)
5             If no node w ∈ M is marked with δ(l(v), aᵢ), create and add it to M;
              Let v become a child of w;
          }
          T := M; i := i + 1;
      } (* End of SCS *)
10
      Procedure ReduceStacks (RS) {
          M := ∅; For (q, q') ∈ Q × (Q \ {q₀}) Do Flag F_{q,q'} := False;
          While ∃v ∈ T Do { (* Process all paths ending in v *)
              T := T \ {v};
```

15   **If** $Action(l(v), a_i) = Reduce(A \rightarrow \alpha)$ **Then** { (* Reduce paths *)
       If $h \geq |\alpha|$ **And** $T \cup M = \emptyset$ **Then** {
          Replace the $|\alpha|$ symbols in front of a_i by the single symbol A;
          If $i \neq i'$ **Then** { $i' := i$; ClearStack(K); }
          Push($K, A \rightarrow \alpha$); $h := h - |\alpha| + 1$;
       } **Else** $h := 0$;
       $R := \{v\}$; $j := |\alpha|$;
       While $j > 0$ **And** $R \neq \emptyset$ **Do** { (* Shorten all paths by at most $|\alpha|$ *)
          $T := T \cup \{(v, j) \mid v \in R$ is a leaf$\}$;
          $R' := \{v \mid v$ is a child of some node in $R\}$;
25   Remove all connecting edges between nodes in R and R';
          $R := R'$; $j := j - 1$;
       }
       For $(w, j) \in T$ **Do** { (* Generate new short paths *)
          For $q \in \delta^{-1}(l(w), j)$ **Do If Not** Flag $F_{q, \delta(q, A)}$ **Then** {
30   Flag $F_{q, \delta(q, A)} :=$ True;
             Create a new node v labelled with q and add v to R;
          }
          Remove node w;
       }
35   **For** $w \in R$ **Do** { (* Add the new state to all paths *)
          Let w become the child of a new root node v labelled with $\delta(l(w), A)$;
          If $\exists v' \in T \cup M : l(v') = \delta(l(w), A)$ **Then**
             MergeTrees(v', v);
          Else $T := T \cup \{v\}$;
40   }
    } **Else If** $Action(q, a_i) = Error$ **Then** {
       Remove the tree rooted in v completely from the graph;
    } **Else** (i.e., $Action(q, a_i) = Shift$) $M := M \cup \{v\}$;
    }
45 $T := M$;
  } (* End of RS *)

    $h := 0$; ClearStack(K); (* Beginning of Main Program *)
    $i := 1$; $T := \emptyset$;
50 **For** each $q \in Q$ with $Action(q, a_1) = Shift$ **Do**
       Create a new node v labelled with q and add v to T;
    While $i < n$ **And** $T \neq \emptyset$ **Do** { SCS; RS; }
    If $T = \emptyset$ **Then** {
       While Not Empty(K) **Do** {
          Pop the top element $A \rightarrow \alpha$ from K;
          Replace the single symbol A in front of a_i by α;
       }
       Return($i - 1$);
59 } **Else** Return(n); (* End of Main Program *)

4 Correctness

We now prove the correctness of the algorithm.

Lemma 1 *The above assured properties of the graph (i.e., it consists of trees, each path consists of at least two nodes, the children of each node are differently labelled, and so are the root nodes contained in T) always hold before and after executing SCS or RS. They also hold whenever line 13 is reached (i.e., the main loop of RS is about to be executed) except for the fact that the root nodes are then contained in $T \cup M$ and not in T alone.*

Proof. By an easy analysis of SCS and RS combined with a simple induction. □

For the next lemma, we need a definition that describes the history of a path during the execution of the algorithm. More precisely, we inductively define an $(m_1, \ldots, m_k)$–*path* as follows, where $m_1, \ldots, m_k \in \{S\} \cup P$. An (ε)–path is a path generated during the **For**–loop of the main program. An $(m_1, \ldots, m_k, S)$–path is an $(m_1, \ldots, m_k)$–path lenghtened by one node due to one execution of SCS. Finally, all paths resulting from an $(m_1, \ldots, m_k)$–path p due to the simulation of a reduction corresponding to a rule $A \to \alpha$ are $(m_1, \ldots, m_k, A \to \alpha)$–paths. These paths are generated (possibly among others) during one execution of the lines 21–40. Clearly, this implies that the ending node v of p satisfies the condition in line 15. Also, note that m_1 is always equal to S because SCS is the first procedure being called.

Lemma 2 *Let p be an $(m_1, \ldots, m_r)$–path and assume SCS has been executed i times. Then there exists a partial stack s such that*

$$(s, a_1 \ldots a_n) \mathop{\vdash}^{m_1} \ldots \mathop{\vdash}^{m_r} (l(p), a_{i+1} \ldots a_n) \ .$$

Proof. If $r = 0$, then $i = 0$ and we can choose $s := l(p)$. For the induction step $r \to r + 1$, let p' be the $(m_1, \ldots, m_r)$–path from which p has been constructed. From the induction hypothesis we know there exists a partial stack s' such that $(s', a_1 \ldots a_n) \mathop{\vdash}^{m_1} \ldots \mathop{\vdash}^{m_r} (l(p'), a_{i'+1} \ldots a_n)$, where i' is the number of executions of SCS before the creation of p.

We first assume $m_{r+1} = S$. Then p has been created due to a call to SCS and p' existed before the execution of this procedure. Let v be the ending node of p. Obviously, from the **For**–loop in the main program and from the management of M in RS, $Action(l(v), a_{i'+1}) = Shift$. Hence,

$$(l(p'), a_{i'+1} \ldots a_n) \mathop{\vdash}^{S} (l(p')\delta(l(v), a_{i'+1}), a_{i'+2} \ldots a_n) \ .$$

Clearly, from the way SCS works, $l(p) = l(p')\delta(l(v), a_{i'+1})$. Since $i = i' + 1$, we thus obtain our desired result by choosing $s := s'$.

We now consider the case $m_{r+1} = A \to \alpha$. Before continuing, we recall the following property of an LR–DFA:

Lemma 3 *Let $q \in Q$, $[A \rightarrow X_1 \ldots X_i \bullet X_{i+1} \ldots X_n, \{w\}] \in q$, $j \leq i$, and let $q' \in \delta^{-1}(q, j)$. Then $[A \rightarrow X_1 \ldots X_{i-j} \bullet X_{i-j+1} \ldots X_n, \{w\}] \in q'$ and*

$$\delta(q', X_{i-j+1} \ldots X_i) = q .$$

Proof. Immediate from the definition of δ^{-1} and the construction of an LR–DFA.   □

Now let $j := |p'|$, and let $v_1, \ldots, v_{j+1}$ be the nodes of p'. From the definition of p, v_{j+1} has fulfilled the condition in line 15, i.e., $Action(l(v_{j+1}), a_{i'+1})$ is equal to $Reduce(A \rightarrow \alpha)$ and therefore $[A \rightarrow \alpha \bullet, \{a_{i'+1}\}] \in l(v_{j+1})$. Assuming $j \geq |\alpha|$, we know that only p is generated from p', labelled with

$$l(p) = l(v_1) \ldots l(v_{j+1-|\alpha|}) \delta(l(v_{j+1-|\alpha|}), A) .$$

From $Action(l(v_{j+1}), a_{i'+1}) = Reduce(A \rightarrow \alpha)$ we also know that

$$(l(p'), a_{i'+1} \ldots a_n) = (l(v_1) \ldots l(v_{j+1}), a_{i'+1} \ldots a_n)$$
$$\vdash^{A \rightarrow \alpha} (l(v_1) \ldots l(v_{j+1-|\alpha|}) \delta(l(v_{j+1-|\alpha|}), A), a_{i'+1} \ldots a_n) = (l(p), a_{i'+1} \ldots a_n) .$$

Since $i = i'$, we again simply choose $s := s'$. (In Lemma 5, we shall see the importance of preserving s'.)

We now assume $j < |\alpha|$. Then p is a short path and consists of two nodes labelled with q and $\delta(q, A)$, where q is in $\delta^{-1}(l(v_1), |\alpha| - j)$. Assume $\alpha = X_1 \ldots X_{|\alpha|}$. Since $l(v_1) \in \delta^{-1}(l(v_{j+1}), j)$ and $[A \rightarrow \alpha \bullet, \{a_{i'+1}\}] \in l(v_{j+1})$, Lemma 3 shows that $[A \rightarrow X_1 \ldots X_{|\alpha|-j} \bullet X_{|\alpha|-j+1} \ldots X_{|\alpha|}, \{a_{i'+1}\}]$ is contained in $l(v_1)$. Similarly, $q \in \delta^{-1}(l(v_1), |\alpha| - j)$ implies $[A \rightarrow \bullet \alpha, \{a_{i'+1}\}] \in q$. Let $t_m := \delta(q, X_1 \ldots X_{m-1})$, $m = 1, \ldots, |\alpha| - j$. By applying Lemma 3 again, $\delta(t_{|\alpha|-j}, X_{|\alpha|-j}) = \delta(q, X_1 \ldots X_{|\alpha|-j}) = l(v_1)$. Thus $t_1 \ldots t_{|\alpha|-j} l(v_1) \ldots l(v_{j+1})$ is a partial stack. Hence, by the induction hypothesis,

$$(t_1 \ldots t_{|\alpha|-j} s', a_1 \ldots a_n) \vdash^{m_1} \ldots \vdash^{m_r} (t_1 \ldots t_{|\alpha|-j} l(p'), a_{i'+1} \ldots a_n) .$$

(Note that $t_1 \ldots t_{|\alpha|-j} s'$ is a partial stack as well because a parsing step never changes the state at the stack bottom. Thus, the first state of s' is equal to $l(v_1)$.) Since $Action(l(v_{j+1}), a_{i'+1}) = Reduce(A \rightarrow \alpha)$, we also have

$$(t_1 \ldots t_{|\alpha|-j} l(p'), a_{i'+1} \ldots a_n) \vdash^{A \rightarrow \alpha} (t_1 \delta(t_1, A), a_{i'+1} \ldots a_n) .$$

But t_1 is equal to q and thus, $t_1 \delta(t_1, A) = l(p)$. We therefore choose s to be equal to $t_1 \ldots t_{|\alpha|-j} s'$.   □

Lemma 4 *Let $(s_1 \ldots s_r, az)$ be some partial configuration which satisfies the condition $Action(s_r, a) = Shift$. Then there exists a string $x \in \Sigma^*$ and a stack t such that $(q_0, xaz) \vdash^* (ts_1 \ldots s_r, az)$.*

Proof. Let t be a path in the LR–DFA from q_0 to some state $q \in \delta^{-1}(s_1, 1)$ (if $s_1 = q_0$, then let $t := \varepsilon$). Then $ts_1 \ldots s_r$ is a path from q_0 to s_r, i.e., $\tilde{\varphi}(ts_1 \ldots s_r)$ is a viable prefix of the grammar. Since the transition $\delta(s_r, a)$ is defined, we conclude that $\tilde{\varphi}(ts_1 \ldots s_r)a$ is a viable prefix as well. Hence, it is possible to choose two strings x and y such that $S \overset{\text{rm}}{\Longrightarrow}{}^{*} \tilde{\varphi}(ts_1 \ldots s_r)ay \overset{\text{rm}}{\Longrightarrow}{}^{*} xay$. From the way an LR parser works, this implies $(q_0, xay) \vdash^{*} (ts_1 \ldots s_r, ay)$. Thus, since the LR parser only has an one–symbol–lookahead, $(q_0, xaz) \vdash^{*} (ts_1 \ldots s_r, az)$.

$\square$

Lemma 5 *The substring parser algorithm always terminates.*

Proof. Clearly, we only have to show that a call to RS always returns. Assuming the opposite, the outer **While**–loop of RS obviously has to run forever. But Lemma 1 shows that $|T|$ is bounded by $|Q|$ at the beginning, and with each pass of the loop one node is removed from T in line 14. Thus, since line 39 contains the only statement in the loop that adds nodes to T, the part of RS that simulates a reduction for all paths endling in some root node $v \in T$ (lines 21–40) must be executed infinite many times. Since $|T| \leq |Q|$, there is at least one path from which infinite many successor paths are generated. Clearly, from the management of the flags $F_{q,\delta(q,A)}$, the total number of short successor paths is bounded by $|Q|^2 - |Q|$. Let p^0 be the first path such that all further successor paths $p^1, p^2, \ldots$ are not short paths. p^0 itself is an $(m_1, \ldots, m_r)$–path for some $m_1, \ldots, m_r$. By Lemma 2, we have $\exists s \colon (s, a_1 \ldots a_n) \vdash^{m_1} \ldots \vdash^{m_r} (l(p^0), a_{i+1} \ldots a_n)$, where i is the number of previous calls to SCS. By reviewing the corresponding part of the proof of Lemma 2, we then conclude that

$$\exists i \, \exists s \, \forall j \colon (s, a_1 \ldots a_n)$$
$$\vdash^{m_1} \ldots \vdash^{m_r} (l(p^0), a_{i+1} \ldots a_n) \underbrace{\vdash^{R} \ldots \vdash^{R}}_{j \text{ times}} (l(p^j), a_{i+1} \ldots a_n) \ .$$

Since $m_1 = S$, we have $Action(s_r, a_1) = Shift$, where $s = s_1 \ldots s_r$. Applying Lemma 4 yields

$$\exists x \, \exists t \colon (q_0, xa_1 \ldots a_n) \vdash^{*} (ts, a_1 \ldots a_n)$$
$$\vdash^{m_1} \ldots \vdash^{m_r} (tl(p^0), a_{i+1} \ldots a_n) \vdash^{R} \ldots \vdash^{R} \ldots \ .$$

Thus, with the input string $xa_1 \ldots a_n$, the original LR parser would run forever as well. This contradiction proves the lemma.

$\square$

Lemma 6 *If* $(q_0, xa_1 \ldots a_n) \vdash^{*} (s_1 \ldots s_j, a_{i+1} \ldots a_n)$ *for some* $x \in \Sigma^*$ *and* $i \in \{1, \ldots, n-1\}$*, then there occurs a path p during the i-th execution of RS such that p is labelled with a suffix of $s_1 \ldots s_j$. More precisely, p is contained in the graph when reaching line 13 at some point of time.*

Proof. Let r denote the number of parsing steps after shifting a_1. If $r = 0$, then s_{j-1} must be a state with $Action(s_{j-1}, a_1) = Shift$, and s_j must be equal to $\delta(s_{j-1}, a_1)$. Clearly,from the executed program code when reaching line 13 for the first time, there exists a path with the labelling $s_{j-1}s_j$, i.e., this path satisfies the claim. Concerning the induction step $r \to r + 1$, we know that $(q_0, xa_1 \ldots a_n) \vdash^{*} (t_1 \ldots t_l, a_{i'+1} \ldots a_n) \vdash (s_1 \ldots s_j, a_{i+1} \ldots a_n)$, and there exists a path p' when reaching line 13 at some point of time during the i'–th execution of RS, where $l(p') = t_f \ldots t_l$ for some f. Let $v \in T$ be the last node of p'. Lemma 5 implies that v is eventually removed from T and thus, p is processed at this point of time. Let us assume that the $(r + 1)$–th parsing step is a reduction that corresponds to a rule $A \rightarrow \alpha$. Then $Action(t_l, a_{i'+1}) = Reduce(A \rightarrow \alpha)$, $s_1 \ldots s_j = t_1 \ldots t_{l-|\alpha|}\delta(t_{l-|\alpha|}, A)$ and $i = i'$. The first of these facts implies that if $|\alpha| \leq l - f$, then p' is replaced by a new path p labelled with $t_f \ldots t_{l-|\alpha|}\delta(t_{l-|\alpha|}, A)$, thus p satisfies the claim. Before reaching line 13 again, a call to MergeTrees possibly replaces p by some shorter path labelled with a suffix of $l(p)$, but then the claim still holds. Now let us assume $|\alpha| > l - f$. Then $(v, |\alpha| - l + f)$ is contained in T when reaching line 28, where v is the first node of p'. Since $t_1 \ldots t_l$ represents a path in the LR–DFA, $t_{l-|\alpha|} \in \delta^{-1}(t_f, |\alpha| - l + f)$. Thus, from the management of the flags $F_{q,\delta(q,A)}$, either now a new short path p with $l(p) = t_{l-|\alpha|}\delta(t_{l-|\alpha|}, A)$ is generated, or such a path has already been generated during the current execution of RS some time before. In both cases, the claim again holds.

We now consider the case that the last parsing step is a shift operation. Then $Action(t_l, a_{i'+1}) = Shift$, $s_1 \ldots s_j = t_1 \ldots t_l\delta(t_l, a_{i'+1})$ and $i = i' + 1$. Clearly, from the management of the set M, $v \in T$ when RS returns. Thus, SCS and RS are both called again. Since SCS converts p' into a path p labelled with $t_f \ldots t_l\delta(t_l, a_{i'+1})$, the claim is again correct because p is still present when entering RS. This completes the proof of the induction step. $\square$

Theorem 1. *The substring parser algorithm is correct.*

Proof. Let $i \in \{2, \ldots, n\}$. We first show the following equivalence:

$$a_1 \ldots a_i \in SS(G)$$
$$\Longleftrightarrow \text{There exists a path (i.e., } T \neq \emptyset \text{) after the } (i - 1)\text{–th execution of RS.}$$

If. Let p be an $(m_1, \ldots, m_r)$–path after the $(i - 1)$–th execution of RS. By Lemma 2, we know that

$$\exists s_1 \ldots s_r : (s_1 \ldots s_r, a_1 \ldots a_n) \stackrel{m_1}{\vdash} \ldots \stackrel{m_r}{\vdash} (l(v_1) \ldots l(v_t), a_i \ldots a_n) ,$$

where $v_1, \ldots, v_t$ are the nodes of p. Since $v_t \in T$ must satisfy the condition $Action(l(v_t), a_i) = Shift$ when leaving RS, we obtain

$$\exists x \, \exists t : (q_0, xa_1 \ldots a_n) \vdash^* (ts_1 \ldots s_r, a_1 \ldots a_n)$$
$$\stackrel{m_1}{\vdash} \ldots \stackrel{m_r}{\vdash} (tl(p), a_i \ldots a_n) \stackrel{S}{\vdash} \ldots$$

by applying Lemma 4. Hence, the LR parser successfully reads the prefix $xa_1 \ldots a_i$. But an LR parser never shifts an errornous symbol. Thus, $xa_1 \ldots a_i$ is the prefix of some sentence $xa_1 \ldots a_i y$. Hence, $a_1 \ldots a_i \in SS(G)$.

Only If. Let $a_1 \ldots a_i \in SS(G)$, i.e., there exist two strings x and y such that $xa_1 \ldots a_i y \in L(G)$. Thus, there also exist two stacks s, t such that

$$(q_0, xa_1 \ldots a_i y) \vdash^* (t, a_i y) \stackrel{S}{\vdash} (s, y) \vdash^* \ldots$$

We also have $(q_0, xa_1 \ldots a_n) \vdash^* (t, a_i \ldots a_n) \vdash^{\underline{S}} (s, a_{i+1} \ldots a_n)$ because the LR parser only has an one–symbol–lookahead. By Lemma 6, there occurs a path p during the $(i-1)$–th execution of RS such that $l(p)$ is a suffix of t. In particular, the last node v of p is labelled with the last state of t. Therefore, v fulfills the condition $Action(l(v), a_i) = Shift$ and thus, p (or a suffix of it due to some call to $MergeTrees$) is still present when returning from RS.

Clearly, by an easy analysis of the main program, the proven equivalence shows that our substring parser works correctly.   □

5   Complexity

We now show that our algorithm runs in linear time relating to the length of the calculated prefix of the input string.

Lemma 7 *Let $C := \{k \mid \exists k', k'' : k' \vdash^{\underline{S}} k'' \vdash^* k\}$ be the set of partial configurations that result from shifting at least one symbol. Then there exists a constant K_1 such that the LR parser, starting with any partial configuration contained in C, never performs K_1 consecutive reductions without decreasing the initial number of elements on the stack.*

Proof. Let $k = (s, y) \in C$. From the definition of C, there exist $k' = (t', ay)$ and $k'' = (t'', y)$ such that $k' \vdash^{\underline{S}} k'' \vdash^* k$. Moreover, when starting with k, let $\tau(k)$ denote the number of reductions performed by the LR parser until either the next symbol is shifted, the stack would contain less than $|s|$ states after the next reduction, or an error occurs. $\tau(k)$ must be a finite number because otherwise, by Lemma 4,

$$\exists x \, \exists t : (q_0, xay) \vdash^* (tt', ay) \vdash^{\underline{S}} (tt'', y) \vdash^* (ts, y) \vdash^{R} \ldots \vdash^{R} \ldots .$$

Hence, in contrast to its properties, the LR parser would never terminate when parsing the string xay.

Now let $s = s_1 \ldots s_r$. Since a reduction corresponding to a rule $A \to \alpha$ at first removes $|\alpha|$ states from the stack and then pushes a new state on its top, none of the $\tau(k)$ reductions discards any of the states $s_1, \ldots, s_{r-1}$ during its removal phase because otherwise, the resulting stack would be shorter than the initial one. In particular, $s_1, \ldots, s_{r-2}$ are never used for determining the pushed states. Thus, only s_{r-1} and s_r have an influence over $\tau(k)$. Furthermore, except for the first symbol of y, $\tau(k)$ is not affected by y either because the LR parser only has an one–symbol–lookahead. Hence, we can choose

$$K_1 := \max\{\tau(k) \mid k \in C \cap ((Q \cup Q^2) \times \Sigma)\} + 1 .$$

The set $(Q \cup Q^2) \times \Sigma$ is finite, and thus, so is K_1.   □

Lemma 8 *Let $I := V \setminus (T \cup M)$ denote the set of internal nodes when reaching the start of the outer **While**–loop of RS (line 13) at some point of time. Then at least one node is removed from I after at most $K_2 := |Q| \cdot K_1 + 1$ additional executions of this loop.*

Proof. Let us first assume that a complete tree is deleted (line 41) during the next K_2 executions. This implies that at least one path is completely removed from the graph. Since the first node of a path is always internal, the lemma is proven in this case.

By Lemma 1, $|T| \leq |Q|$ and $|M| \leq |Q|$ and thus, the condition in line 43 cannot be satisfied more than $|Q|$ times. Hence, at least $K_2 - |Q| = |Q|(K_1 - 1) + 1$ reduction simulations are preformed. From the bound on T, there is some path p with at least K_1 such simulations. Thus, by Lemma 7, p must have been shortened in the meantime, and this implies that at least one node $v \in I$ must have been deleted.   □

Lemma 9 *The difference in the number of internal nodes before and after executing RS is bounded by a constant.*

Proof. Normally, the simulation of a reduction removes more nodes than generating new ones. The only reductions that increase the length of a path by one node are those corresponding to ε–*rules*, e.g. $A \to \varepsilon$. Thus, at most $K_1 - 1$ such nodes can be found at the end of each path because, as already seen, a path is getting shorter after at most K_1 reductions. From the structure of the graph when leaving RS, the ending nodes of all paths are contained in T, the children of the ending nodes are children of the nodes in T, and so on. By Lemma 1, not only T, but also any set of children cannot contain more than $|Q|$ nodes. Thus the number of the above mentioned nodes is bounded by $|Q| + |Q|^2 + \cdots + |Q|^{K_1 - 1}$. The at most $|Q|$ nodes in T are not internal ones, but the former root nodes in T before executing RS may now be internal, thus the given number is also a bound for the internal nodes. Hence, from the fact that the generation of the short paths additionally produces at most $2(|Q|^2 - |Q|)$ nodes, the lemma is proven.   □

Theorem 2. *Let $j := \max\{i \mid a_1 \ldots a_i \in SS(G)\}$. Then the time consumed by the algorithm is bounded by $O(j)$. In particular, the algorithm takes at most $O(n)$ time.*

Proof. SCS and RS are called for $j - 1$ times. Clearly, a call to SCS increases the number of internal nodes by at most $|Q|$. Together with Lemma 9, this implies a total $O(j)$ bound on the number of new internal nodes after returning from SCS and RS. Thus, by Lemma 8, the number of executions of the outer **While**–loop of RS is also bounded by $O(j)$. Recalling the $|Q|$ bound on $|T|$ and on the number of children of any node, it is easy to see that one instance of this loop can be executed in constant time, ignoring the costs of deleting and merging trees (lines 38 and 42). Clearly, a call to SCS only takes some constant time as well. Thus, there is an $O(j)$ bound on the total time consumed by the algorithm (still ignoring lines 38 and 42) and therefore, we also have a linear bound on the total number of all generated nodes. This implies that the deletion of all trees can be done in $O(j)$ time, and moreover, the number of single calls to *MergeTrees*(v_1, v_2) cannot exceed $O(j)$ because such a call removes at least the node v_2. This completes the proof on the linear time bound of our algorithm.   □

6   Condensation of Substrings

During the parsing process, an LR parser converts an input word $w \in L(G)$ step by step into the sentential forms that occur when generating the rightmost derivation of w. More precisely, if $(q_0, xy) \vdash^* (s, y)$, then $\tilde{\varphi}(s)y$ is such a sentential form, and it is possible to derive x from $\tilde{\varphi}(s)$. Similarly, the substring parser is able to convert the already read prefix x of an input string w into a corresponding *partial sentential form* β, i.e., $\beta \overset{rm}{\Longrightarrow}^* x$, and there exist some α, γ such that $\alpha\beta\gamma \in S(G)$. This *condensation* of x is advantageous if x must be parsed again later. For example, if $x = abcde$ and there are two productions $A \to bc$ and $B \to Ad$ which must always be applied to derive x, then the partial sentential form aBe can be processed more quickly than $abcde$ because two reductions are already done. Thus, it is useful to replace a substring with its condensation whenever possible. Of course, this can only be done if the "stored" reductions are always applied to the substring. We therefore restrict the conversion of x into β by the condition $\forall v, z\colon vxz \in L(G) \implies v\beta z \in S(G)$. Thus, we do not want a partial sentential form to depend on some special context strings v and z.

The condensation feature is realized by the program lines with a bar ($|$) on their left sides. The algorithm as before returns an index j corresponding to the calculated prefix $a_1 \dots a_j$, but moreover, this prefix is now replaced by an appropriate condensation of it. The idea which supports the additional lines is as follows. Let $P^j(G) := \{v \mid \exists z\colon va_1 \dots a_j z \in L(G)\}$, and let us assume that the following *C-condition* holds for some indices i, j with $1 \leq i \leq j \leq n$:

$$\exists s_1, \dots, s_r \in Q \ \forall v \in P^{j-1}(G) \ \exists t \in Q^* \ \exists k_1, \dots, k_l \in (\{ts_1\}Q^+) \times \Sigma^*\colon$$
$$(q_0, va_1 \dots a_n) \vdash^* (ts_1, a_i \dots a_n) \vdash k_1 \vdash \dots \vdash k_l = (ts_1 \dots s_r, a_j \dots a_n).$$

Since $k_1, \dots, k_l \in (\{ts_1\}Q^+) \times \Sigma^*$, it can be easily seen that beginning with $(ts_1, a_i \dots a_n)$, the state s_1 is never removed from the stack. Therefore these transitions only depend on the common state s_1 and not on t. Thus, from the way an LR parser works, when parsing a string $va_1 \dots a_j z \in L(G)$, the substring $a_i \dots a_{j-1}$ is always reduced to $\varphi(s_1 \dots s_r)$. (Recall that the parser cannot look beyond the symbol a_j because of its one–symbol–lookahead.) Hence, $va_1 \dots a_{i-1}\varphi(s_1 \dots s_r)a_j z \in S(G)$, and therefore $a_1 \dots a_{i-1}\varphi(s_1 \dots s_r)a_j$ is a condensation of $a_1 \dots a_j$.

The condensation algorithm detects whether the above *C-condition* holds, and, if this is the case, changes the input string appropriately by replacing $a_i \dots a_{j-1}$ with $\varphi(s_1 \dots s_r)$. This replacement is done step by step by applying the corresponding reductions to the input list. The following two lemmas show how the detection is managed. The first one implies a possibility to test whether the C-condition is true and is a refinement of Lemma 6.

Lemma 10 *Let $i \in \{2, \dots, n\}$, and let $v \in \Sigma^*$ and $y = a_1 \dots a_n \in \Sigma^{\geq 2}$ be two strings such that*

$$(q_0, va_1 \dots a_n) \vdash^* (t, a_{i-1} \dots a_n) \overset{S}{\vdash} k_1 \vdash \dots \vdash k_l \overset{S}{\vdash} (t', a_{i+1} \dots a_n) \;,$$

where $k_j = (s^j, a_i \ldots a_n)$, for $j \in \{1, \ldots, l\}$. Then there exists a path p in the graph and an index $j \in \{1, \ldots, l\}$ whenever reaching line 13 during the $(i-1)$-th execution of RS such that $l(p)$ is a suffix of s^j.

Proof. Let f denote the total number of executions of the outer **While**–loop of RS, and for $b \in \{1, \ldots, f+1\}$, let G_b be the graph and p_b the corresponding path when reaching line 13 for the b-th time. We prove the lemma by an induction on b. Clearly, G_1 only consists of paths that result from shifting a_{i-1}. From the proof of correctness, we know that the lemma holds in this case with $j = 1$. Now let us consider the induction step $b \to b+1$. By the induction hypothesis, there exists a path p_b and an index j_b satisfying the lemma. Let v be the ending node of p_b. During the next execution of the outer **While**–loop, possibly p_b is not considered. Then we can choose $p_{b+1} := p_b$ and $j_{b+1} := j_b$. If p_b is replaced by another path due to the simulation of a reduction, then we can choose this path as p_{b+1} and $j_{b+1} := j_b + 1$ because $l(p_{b+1})$ must be a suffix of s^{j_b+1}. Note that in both cases, a call to *MergeTrees* possibly deletes p_{b+1}, but then there must be a path that is labelled with a suffix of $l(p_{b+1})$, and thus we can then choose this shorter path as p_{b+1}.

Finally, we consider the case that p_b is completely removed from the graph and there does not exist a suitable path for satisfying the lemma afterwards. This cannot result from an execution of line 42 because $Action(l(v), a_i)$ is either *Shift* or *Reduce*. Thus, this case is only possible if the generation of an appropriate new short path p_{b+1} fails due to the violation of the condition in line 29, where the two states of p_{b+1} match the last two states of s^{j_b+1}. But then p_{b+1} must have been created some time before, i.e., there exists some $c \in \{1, \ldots, b\}$ such that p_{b+1} existed in the graph before executing the main loop for the c-th time. Now we can restart the induction, beginning with $b = c$. Since this case is only possible for at most $|Q|^2 - |Q|$ times, we finally succeed in proving the existance of p_{b+1} and j_{b+1}. □

Let us now assume that SCS has been called $i-1$ times, and there only exists one root node w in the graph when reaching line 13 at some point of time. Then, by Lemma 10, for every $v \in P^i(G)$ there is some path p in the graph such that $\exists t' \in Q^* : (q_0, va_1 \ldots a_i) \vdash^* (t'l(p), a_i)$. (Note that a_i will be shifted some time later because $va_1 \ldots a_i$ is a prefix of a sentence of G and the parser only has an one–symbol–lookahead.) But every path in the graph ends with $s_1 := l(w)$, and thus we know that the following special case of the C–condition is true whenever there is only one root node:

$$\exists s_1 \in Q \; \forall v \in P^i(G) \; \exists t \in Q^* : (q_0, va_1 \ldots a_i) \vdash^* (ts_1, a_i) \ .$$

From the discussion at the beginning of this section, it is easy to see that the C–condition remains valid as long as the common state s_1 is not popped off the stack, i.e., as long as the node w is not removed from the graph. From the way the graph is maintained, it then follows that all paths will always end in a common part that starts with w, and only this part is altered during this time. We now show that the length of this common path is always contained in the variable h.

Lemma 11 *When there is only one root node v in the graph, the variable h contains the length of the common suffix of all paths. If there is more than one root node, then h is equal to zero.*

Proof. h must be set to zero (line 48) while initializing the graph (lines 50–51) because the graph only consists of single nodes. When calling SCS, h must be increased by one in the case that there is only one root node because this procedure then appends a new node to it. Otherwise, h must remain zero even if there is only one root node afterwards (line 3). Now we consider a call to RS. Clearly, h must be changed during an execution of the outer **While**-loop only if the condition in line 15 is true. Since a reduction corresponding to a rule $A \rightarrow \alpha$ first removes $|\alpha|$ nodes from a path and then creates a new one, h must be decreased by $|\alpha| - 1$ (line 19), but only if $|\alpha| \leq h$ (line 16). Otherwise, there does not exist a common path any more after the removal of the $|\alpha|$ nodes, and thus h must be reset to zero (line 20). Note that the condition $T \cup M = \emptyset$ in line 16, which is used to test whether there is exactly one root node in the graph, is correct because one node has been removed from T before (see line 14).   □

We now explain some more of the remaining lines of our condensation algorithm. We already know that as long as there is only one root node in the graph, all reductions that are made in the meantime are common to all paths and thus they can be applied to the input string as well. For one single rule $A \rightarrow \alpha$, the corresponding modifications of the input string are done in line 17. In order to implement these modifications efficiently, the input string should be administered as a double–chained–list.

As we have mentioned earlier, the common reductions that are applied to the input string lead to a valid condensation, but only if the next input symbol is shifted some time later. When using a canonical LR parser, we do not have to care about this problem because such a parser is known to never perform a reduction if the next symbol causes an error. But non–canonical parsers (e.g. SLR parsers) may still perform some wrong reductions before detecting the error situation. Since the substring parser has no possibility to detect in advance whether the next symbol can be shifted or not, it stores every reduction from the last shift operation on in a stack K such that these reductions can be undone when necessary. More precisely, the stack K is maintained in the following way. At the beginning, K is cleared in line 48. Then all reductions that affect the input list are pushed onto K in line 19. Since all these reductions are definitely correct when the next symbol is shifted, they never need to be undone in this case and thus, they can be discarded from K. This is guaranteed by the management of the additional index i' in line 18. The additional **While**-loop in lines 54–57 restores the part of the list that was changed during the last reductions. Clearly, this is only necessary if the condition in line 53 is true because otherwise the complete input string has been successfully shifted and therefore all reductions are correct. It is easy to see that the loop pops one reduction after another from K and restores the list step by step. Therefore, the finally calculated condensation is always correct. Note that the rest of the algorithm is not affected by the condensation part because it has no influence over the graph structure. Furthermore, from the time complexity of the original algorithm, we know that the total number of rules pushed onto K cannot exceed $O(j)$, where $a_1 \ldots a_j$ is the calculated prefix of the input string. The time consumed by the **While**–loop in lines 54–57 is therefore as well bounded by $O(j)$. Since the additional lines 16–20 only take some constant time, the total $O(j)$ time bound of the algorithm is still valid.

When implementing the program, recall that the stack K is not necessary if the substring parser uses a canonical LR(1)–DFA.

7 Recognizing Substrings of LR(k) Languages

In this section, we first show that with some minor modifications, our substring parser can even be used with canonical LR(k) languages, where $k > 1$. Later, we also discuss the case of non–canonical LR(k) languages. Canonical LR(k) languages are easier to handle because their parsers have the *valid–prefix property*, i.e., after parsing the prefix x of some input string xy and assuming that q is the current state on the stack top, $Action(q, y_k) \neq Error$ iff xy_k is the prefix of some sentence in $L(G)$, where y_k denotes the first k symbols of y. In the context of our substring recognizing problem, this means that when our algorithm returns, the next $k - 1$ unshifted symbols also belong to the longest prefix in $SS(G)$. (The next k unshifted symbols do not belong to it because they were responsible for the *Action* function to fail.)

The modified algorithm works as follows. Let $w = a_1 \ldots a_n$ be the input string. At first, the maximum prefix x of $a_1 \ldots a_{k-1}$ is determined such that $x \in SS(G)$. If $|x| < k - 1$ (or $|w| < k$), then clearly there is nothing left to do. Otherwise, we start our previous algorithm with three modifications. Firstly, wherever the function *Action* is used, the next k symbols must be used for the lookahead string. Secondly, the condition $i < n$ in line 52 must be replaced by $i < n - k + 1$. And finally, the returned index must be increased by $k - 1$. This leads to the following algorithm.

$j := 1;$
While $j < k$ **And** $j \leq n$ **And** $a_1 \ldots a_j \in SS(G) \cap \Sigma^j$ **Do** $j := j + 1;$
If $j = k$ **And** $n \geq k$ **Then** {
    Execute the previous algorithm with the above mentioned modifications;
    Return($j + k - 1$); (* where j is returned by the original algorithm *)
} **Else** Return($j - 1$);

The correctness and the linear complexity can be proven in nearly the same way as before. Note that the sets $SS(G) \cap \Sigma^i$, for $i \in \{1, \ldots, k - 1\}$, can be precalculated from the canonical LR(k)–DFA $(Q, \Gamma, \delta, q_0, F)$ of G by using the equation

$$SS(G) \cap \Sigma^i := \{w \in \Sigma^i | \exists q \in Q \exists [A \to \alpha_\bullet \beta, \{x\}] \in q \exists y, z \in \Sigma^* : x = ywz\} .$$

The new algorithm is *not* correct when using LR–DFAs of non–canonical LR(k) parsers, e.g. SLR(k) parsers, because these parsers do not have the valid–prefix property. In fact, after the modified original algorithm returns with an index j, we only know that $a_1 \ldots a_j \in SS(G)$ and $a_1 \ldots a_{j+k} \notin SS(G)$. However, we are able to present a solution to this problem even in this case.

Clearly, for every $w \in SS(G)$ there exists a string $y \in \Sigma^{\leq k-1}$ such that either $y \in \Sigma^{k-1}$ and still $wy \in SS(G)$, or $y \in \Sigma^{<k-1}$ and wy is a suffix of a sentence $xwy \in L(G)$. The idea now is to apply the modified algorithm to the new input string $w' := wy$. Note that if $y \in \Sigma^{k-1}$, then the algorithm accepts at least the prefix w because the lookahead of the substring parser always contains a substring of wy as long as the last symbol of w has not yet been shifted, and this will never happen due to the modified condition $i < n - k + 1$ in line 52. In the other case, i.e., $y \in \Sigma^{<k-1}$, the substring parser completely accepts wy because otherwise the corresponding original LR parser would refuse to accept sentences of G that end with wy. Thus the returned pointer j corresponds to the last symbol of w or to one symbol of y iff $w \in SS(G)$. Unfortunately, since y is unknown, we in general have to perform this test for all strings $y \in \Sigma^{\leq k-1}$.

By using this method, we can check whether $w := a_1 \ldots a_{j+\lfloor k/2 \rfloor}$ is a substring or not. Clearly, with additional $O(\log k)$ interval halvings we can then determine the exact solution. Note that while the complexity of this algorithm is still $O(j)$, there are possibly $O(2^k \log k)$ executions of the original algorithm, and thus this method does not seem to be practical if k is not small.

8   Conclusion

We have presented a linear–time bounded algorithm for recognizing and condensing substrings of LR(k) languages. Practical experience has shown that this substring parser is nearly as fast as the corresponding normal LR parser. The substring parser has primarily been developed in order to generate a new algorithm for syntax error correction and recovery. This algorithm depends on the ideas of Richter [5] and divides an incorrect program into several parts such that on one hand each part contains at least one syntax error, but on the other hand a shorter substring of any part does not contain any syntax errors. This is easily done by firstly determining the longest error–free prefix of the program, and then secondly using the substring parser on the rest of the program to calculate the next part. The second step is then repeated until the complete program is analysed. Usually, the syntax errors can be found at the borders of the parts, and contrary to Richter's opinion, numerous tests with sample programs that contained many different errors have shown that it is possible to obtain very good corrections by using the substring parser on three or more successive parts, where one part contains a trial correction and the other ones supply some context information. The length of the determined prefix then represents the quality of the tested correction. By condensing the different parts with the extension presented in Section 4, it is even possible to give the programmer an overview of the structure of his program. For example, running the substring parser with a standard PASCAL grammar and the input string

If $i = 1$ **Then** $i := i + 1$; **Else** $i := 0$;

generates the following output:

> **If** *BooleanCondition* **Then** *Statement* ;

Since this is the longest prefix of the input string that represents a substring of some correct PASCAL program, the programmer knows that an error occurs when appending the keyword **Else** (a semicolon followed by **Else** is incorrect). Also, the condition and the statement which are both meaningless in this special form are replaced by their more abstract grammar variables.

The resulting algorithm is fast and has several advantages over other correction methods, e.g. the advantage of never detecting spurious errors. Details will be published elsewhere.

A   The Problems in the Substring Parser of Bates and Lavie

In [2], Bates and Lavie also present a substring parser for LR grammars. But unfortunately, both the correctness and the linear complexity are proven incorrectly, as we shall now demonstrate (familiarity with [2] is assumed).

We again use the SLR(1) grammar given in the second section as an example. Let $c = (q_3, b)$ be a partial configuration. Since there is only one path from q_0 to q_3 in Fig. 1, there exists as well only one configuration such that c is an inner part of it. This configuration is $(q_0 q_1 q_2 q_3, b)$ and results from shifting the symbol "a" for three times:

$$(q_0, aaab) \vdash^S (q_0 q_1, aab) \vdash^S (q_0 q_1 q_2, ab) \vdash^S (q_0 q_1 q_2 q_3, b) \ .$$

Clearly, the next configuration then results from a reduction corresponding to the rule $A \to aa$:

$$(q_0 q_1 q_2 q_3, b) \vdash^{A \to aa} (q_0 q_1 q_5, b) \ .$$

By using the notions introduced in [2], these facts can be written as

$$c = ([q_3], b, 1) \ , \quad M(c) = ([q_0, q_1, q_2, q_3], aaab, 4) \ ,$$
$$next(M(c)) = \{([q_0, q_1, q_5], aaab, 4)\} \ .$$

Now we determine the set $next(c)$. Clearly, $\mathrm{LONG}(A) = \{q_4, q_5\}$. Since the right hand side of the rule $A \to aa$ is longer than the current stack in c, we conclude that $next(c)$ results from a long reduction. By Definition 6 in [2], we have $next(c) = \{([q_4], b, 1), ([q_5], b, 1)\}$. There again is only one path from q_0 to q_4, namely $q_0 q_4$, and only one path from q_0 to q_5, namely $q_0 q_1 q_5$. Moreover, the corresponding configurations may only result from the following parsing steps:

$$(q_0, aab) \vdash^S (q_0 q_1, ab) \vdash^S (q_0 q_1 q_2, b) \vdash^{A \to aa} (q_0 q_4, b)$$

$$(q_0, aaab) \vdash^S (q_0 q_1, aab) \vdash^S (q_0 q_1 q_2, ab) \vdash^S (q_0 q_1 q_2 q_3, b) \vdash^{A \to aa} (q_0 q_1 q_5, b) \ .$$

Thus, we have $M(next(c)) = \{([q_0, q_4], aab, 3), ([q_0, q_1, q_5], aaab, 4)\}$. The *Simulation Lemma (Lemma 6)* claims that $M(next(c)) = next(M(c))$, where C denotes any set of stack configurations. Therefore, with $C := \{c\}$, this lemma is obviously wrong. But then the complete proof of correctness is no longer valid, either.

The complexity analysis given in [2] is correct for grammars without ε-*rules*, i.e., rules of the form $A \to \varepsilon$. But Section 4.2, where the analysis is extended to grammers which include such rules, contains a severe error. Lemma 15 states that in every sentence of any LR grammer G, the number of hidden epsilons between two nonepsilon terminal symbols is always bounded by some constant that only depends on G. But it is possible to present a counterexample. Let G be the following grammar:

$$S \longrightarrow Ab \qquad A \longrightarrow aAB \,|\, c \qquad B \longrightarrow \varepsilon \ .$$

G is LR(0), and it is easy to see that $L(G) = \{a^k c\varepsilon^k b \,|\, k \in \mathbb{N}_0\}$. Clearly, this contradicts Lemma 15. But then the rest of the complexity analysis is also incorrect.

Acknowledgment. I would like to thank Prof. Dr. M. Clausen, F. Kurth and Prof. Dr. N. Blum for helpful comments.

References

[1]   A. V. Aho, J. D. Ullman: *The Theory of Parsing, Translation and Compiling. Vol. 1: Parsing.* Prentice Hall (1972).

[2]   J. Bates, A. Lavie: *Recognizing Substrings of LR(k) Languages in Linear Time.* ACM Transactions on Programming Languages and Systems **16**(3) (1994), 1051–1077.

[3]   G. Clarke, D. T. Barnard: *An LR Substring Parser Applied in a Parallel Environment.* Journal of Parallel and Distributed Computing **35** (1996), 2–17.

[4]   G. V. Cormack: *An LR Substring Parser for Noncorrecting Syntax Error Recovery.* ACM SIGPLAN **24**(7) (1989), 161–169.

[5]   H. Richter: *Noncorrecting Syntax Error Recovery.* ACM Transactions on Programming Languages and Systems **7**(3) (1985), 478–489.

Minimal Cover-Automata for Finite Languages[*]

Cezar Câmpeanu, Nicolae Sântean, and Sheng Yu

Department of Computer Science
University of Western Ontario
London, Ontario, Canada N6A 5B7
{cezar,santean,syu}@csd.uwo.ca

Abstract. A cover-automaton A of a finite language $L \subseteq \Sigma^*$ is a finite automaton that accepts all words in L and possibly other words that are longer than any word in L. A minimal deterministic cover automaton of a finite language L usually has a smaller size than a minimal DFA that accept L. Thus, cover automata can be used to reduce the size of the representations of finite languages in practice. In this paper, we describe an efficient algorithm that, for a given DFA accepting a finite language, constructs a minimal deterministic finite cover- automaton of the language. We also give algorithms for the boolean operations on deterministic cover automata, i.e., on the finite languages they represent.

1 Introduction

Regular languages and finite automata are widely used in many areas such as lexical analysis, string matching, circuit testing, image compression, and parallel processing. However, many applications of regular languages use actually only finite languages. The number of states of a finite automaton that accepts a finite language is at least one more than the length of the longest word in the language, and can even be in the order of exponential to that number. If we do not restrict an automaton to accept the exact given finite language but allow it to accept extra words that are longer than the longest word in the language, we may obtain an automaton such that the number of states is significantly reduced. In most applications, we know what is the maximum length of the words in the language, and the systems usually keep track of the length of an input word anyway. So, for a finite language, we can use such an automaton plus an integer to check the membership of the language. This is the basic idea behind cover automata for finite languages.

Informally, a cover-automaton A of a finite language $L \subseteq \Sigma^*$ is a finite automaton that accepts all words in L and possibly other words that are longer than any word in L. In many cases, a minimal deterministic cover automaton of a finite language L has a much smaller size than a minimal DFA that accept

---

[*] This research is supported by the Natural Sciences and Engineering Research Council of Canada grants OGP0041630.

L. Thus, cover automata can be used to reduce the size of automata for finite languages in practice.

Intuitively, a finite automaton that accepts a finite language (exactly) can be viewed as having structures for the following two functionalities:

1. checking the patterns of the words in the language, and
2. controlling the lengths of the words.

In a high-level programming language environment, the length-control function is much easier to implement by counting with an integer than by using the structures of an automaton. Furthermore, the system usually does the length-counting anyway. Therefore, a DFA accepting a finite language may leave out the structures for the length-control function and, thus, reduce its complexity.

The concept of cover automata is not totally new. Similar concepts have been studied in different contexts and for different purposes. See, for example, [1,7,4,10]. Most of previous work has been in the study of a descriptive complexity measure of arbitrary languages, which is called "automaticity" by Shallit et al. [10]. In our study, we consider cover automata as an implementing method that may reduce the size of the automata that represent finite languages.

In this paper, as our main result, we give an efficient algorithm that, for a given finite language (given as a deterministic finite automaton or a cover automaton), constructs a *minimal* cover automaton for the language. Note that for a given finite language, there might be several minimal cover automata that are not equivalent under a morphism. We will show that, however, they all have the same number of states.

2   Preliminaries

Let T be a set. Then by $\#T$ we mean the cardinality of T. The elements of T^* are called strings or words. The empty string is denoted by λ. If $w \in T^*$ then $|w|$ is the length of x.

We define $T^l = \{w \in T^* \mid |w| = l\}$, $T^{\leq l} = \bigcup_{i=0}^{l} T^i$, and $T^{<l} = \bigcup_{i=0}^{l-1} T^i$. We say that x is a prefix of y, denoted $x \preceq_p y$, if $y = xz$ for some $z \in T^*$. The relation $\preceq_p$ is a partial order on T^*. If $T = \{t_1, \dots, t_k\}$ is an ordered set, $k > 0$, the quasi-lexicographical order on T^*, denoted $\prec$, is defined by: $x \prec y$ iff $|x| < |y|$ or $|x| = |y|$ and $x = zt_iv$, $y = zt_ju$, $i < j$, for some $z, u, v \in T^*$ and $1 \leq i, j \leq k$. Denote $x \preceq y$ if $x \prec y$ or $x = y$.

We say that x is a prefix of y, denoted $x \preceq_p y$, if $y = xz$ for some $z \in T$.

A deterministic finite automaton (DFA) is a quintuple $A = (\Sigma, Q, q_0, \delta, F)$, where Σ and Q are finite nonempty sets, $q_0 \in Q$, $F \subseteq Q$ and $\delta : Q \times \Sigma \longrightarrow Q$ is the transition function. We can extend δ from $Q \times \Sigma$ to $Q \times \Sigma^*$ by

$$\overline{\delta}(s, \lambda) = s$$

$$\overline{\delta}(s, aw) = \overline{\delta}(\delta(s, a), w).$$

We usually denote $\overline{\delta}$ by δ.

The language recognised by the automaton A is $L(A) = \{w \in \Sigma^* \mid \delta(q_0, w) \in F\}$. For simplicity, we assume that $Q = \{0, 1, \ldots, \#Q - 1\}$ and $q_0 = 0$. In what follows we assume that δ is a total function, i.e., the automaton is complete.

Let l be the length of the longest word(s) in the finite language L. A DFA A such that $L(A) \cap \Sigma^{\leq l} = L$ is called a *deterministic finite cover-automaton* (DFCA) of L. Let $A = (Q, \Sigma, \delta, 0, F)$ be a DFCA of a finite language L. We say that A is a *minimal* DFCA of L if for every DFCA $B = (Q', \Sigma, \delta', 0, F')$ of L we have $\#Q \leq \#Q'$.

Let $A = (Q, \Sigma, \delta, 0, F)$ be a DFA. Then

a) $q \in Q$ is said to be accessible if there exists $w \in \Sigma^*$ such that $\delta(0, w) = q$,

b) q is said to be useful (coaccessible) if there exists $w \in \Sigma^*$ such that $\delta(q, w) \in F$.

It is clear that for every DFA A there exists an automaton A' such that $L(A') = L(A)$ and all the states of A' are accessible and at most one of the states is not useful (the sink state). The DFA A' is called a *reduced* DFA.

In what follows we shall use only reduced DFA.

3   Similarity Sequences and Similarity Sets

In this section, we describe the L-similarity relation on Σ^*, which is a generalisation of the equivalence relation $\equiv_L$ ($x \equiv_L y$: $xz \in L$ iff $yz \in L$ for all $z \in \Sigma^*$). The notion of L-similarity was introduced in [7] and studied in [4] etc. In this paper, L-similarity is used to establish our algorithms.

Let Σ be an alphabet, $L \subseteq \Sigma^*$ a finite language, and l the length of the longest word(s) in L. Let $x, y \in \Sigma^*$. We define the following relations:

(1) $x \sim_L y$ if for all $z \in \Sigma^*$ such that $|xz| \leq l$ and $|yz| \leq l$, $xz \in L$ iff $yz \in L$;

(2) $x \not\sim_L y$ if $x \sim_L y$ does not hold.

The relation $\sim_L$ is called *similarity* relation with respect to L.

Note that the relation $\sim_L$ is reflexive, symmetric, but not transitive. For example, let $\Sigma = \{a, b\}$ and $L = \{aab, baa, aabb\}$. It is clear that $aab \sim_L aabb$ and $baa \sim_L aabb$, but $aab \not\sim_L baa$.

The following lemma is obvious:

Lemma 1 *Let $L \subseteq \Sigma^*$ be a finite language and $x, y, z \in \Sigma^*$, $|x| \leq |y| \leq |z|$. The following statements hold:*

1. *If $x \sim_L y$, $x \sim_L z$, then $y \sim_L z$.*
2. *If $x \sim_L y$, $y \sim_L z$, then $x \sim_L z$.*
3. *If $x \sim_L y$, $y \not\sim_L z$, then $x \not\sim_L z$.*

If $x \not\sim_L y$ and $y \sim_L z$, we cannot say anything about the similarity relation between x and z.

Example 1. Let $x, y, z \in \Sigma^*$, $|x| \leq |y| \leq |z|$. We may have

1) $x \not\sim_L y$, $y \sim_L z$ and $x \sim_L z$, or

2) $x \not\sim_L y$, $y \sim_L z$ and $x \not\sim_L z$.

Indeed, if $L = \{aa, aaa, bbb, bbbb, aaab\}$ we have 1) if we choose $x = aa$, $y = bbb$, $z = bbbb$, and 2) if we choose $x = aa$, $y = bba$, $z = abba$.

Definition 1. *Let $L \in \Sigma^*$ be a finite language.*

1. *A set $S \subseteq \Sigma^*$ is called an L-similarity set if $x \sim_L y$ for every pair $x, y \in S$.*
2. *A sequence of words $[x_1, \ldots, x_n]$ over Σ is called a dissimilar sequence of L if $x_i \not\sim_L x_j$ for each pair i, j, $1 \leq i, j \leq n$ and $i \neq j$.*
3. *A dissimilar sequence $[x_1, \ldots, x_n]$ is called a canonical dissimilar sequence of L if there exists a partition $\pi = \{S_1, \ldots, S_n\}$ of Σ^* such that for each i, $1 \leq i \leq n$, $x_i \in S_i$, and S_i is a L-similarity set.*
4. *A dissimilar sequence $[x_1, \ldots, x_n]$ of L is called a maximal dissimilar sequence of L if for any dissimilar sequence $[y_1, \ldots, y_m]$ of L, $m \leq n$.*

Theorem 1. *A dissimilar sequence of L is a canonical dissimilar sequence of L if and only if it is a maximal dissimilar sequence of L.*

Proof. Let L be a finite language. Let $[x_1, \ldots, x_n]$ be a canonical dissimilar sequence of L and $\pi = \{S_1, \ldots, S_n\}$ the corresponding partition of Σ^* such that for each i, $1 \leq i \leq n$, S_i is an L-similarity set. Let $[y_1, \ldots, y_m]$ be an arbitrary dissimilar sequence of L. Assume that $m > n$. Then there are y_i and y_j, $i \neq j$, such that $y_i, y_j \in S_k$ for some k, $1 \leq k \leq n$. Since S_k is a L-similarity set, $y_i \sim_L y_j$. This is a contradiction. Then, the assumption that $m > n$ is false, and we conclude that $[x_1, \ldots, x_n]$ is a maximal dissimilar sequence.

Conversely, let $[x_1, \ldots, x_n]$ a maximal dissimilar sequence of L. Without loss of generality we can suppose that $|x_1| \leq \ldots \leq |x_n|$. For $i = 1, \ldots, n$, define

$$X_i = \{y \in \Sigma^* \mid y \sim_L x_i \text{ and } y \notin X_j \text{ for } j < i\}.$$

Note that for each $y \in \Sigma^*$, $y \sim_L x_i$ for at least one i, $1 \leq i \leq n$, since $[x_1, \ldots, x_n]$ is a *maximal* dissimilar sequence. Thus, $\pi = \{X_1, \ldots, X_n\}$ is a partition of Σ^*. The remaining task of the proof is to show that each X_i, $1 \leq i \leq n$, is a similarity set.

We assume the contrary, i.e., for some i, $1 \leq i \leq n$, there exist $y, z \in X_i$ such that $y \not\sim_L z$. We know that $x_i \sim_L y$ and $x_i \sim_L z$ by the definition of X_i. We have the following three cases: (1) $|x_i| < |y|, |z|$, (2) $|y| \leq |x_i| \leq |z|$ (or $|z| \leq |x_i| \leq |y|$), and (3) $|x_i| > |y|, |z|$. If (1) or (2), then $y \sim_L z$ by Lemma 1. This would contradict our assumption. If (3), then it is easy to prove that $y \not\sim x_j$ and $z \not\sim x_j$, for all $j \neq i$, using Lemma 1 and the definition of X_i. Then we can replace x_i by both y and z to obtain a longer dissimilar sequence $[x_1, \ldots, x_{i-1}, y, z, x_{i+1}, \ldots, x_n]$. This contradicts the fact that $[x_1, \ldots, x_{i-1}, x_i, x_{i+1}, \ldots, x_n]$ is a maximal dissimilar sequence of L. Hence, $y \sim z$ and X_i is a similarity set.

Corollary 1. *For each finite language L, there is a unique number $N(L)$ which is the number of elements in any canonical dissimilar sequence of L.*

Theorem 2. *Let S_1 and S_2 be two L-similarity sets and x_1 and x_2 the shortest words in S_1 and S_2, respectively. If $x_1 \sim_L x_2$ then $S_1 \cup S_2$ is a L-similarity set.*

Proof. It suffices to prove that for an arbitrary word $y_1 \in S_1$ and an arbitrary word $y_2 \in S_2$, $y_1 \sim_L y_2$ holds. Without loss of generality, we assume that $|x_1| \leq |x_2|$. We know that $|x_1| \leq |y_1|$ and $|x_2| \leq |y_2|$. Since $x_1 \sim_L x_2$ and $x_2 \sim_L y_2$, we have $x_1 \sim_L y_2$ (Lemma 1 (2)), and since $x_1 \sim_L y_1$ and $x_1 \sim_L y_2$, we have $y_1 \sim_L y_2$ (Lemma 1 (1)).

4   Similarity Relations on States

Let $A = (Q, \Sigma, \delta, 0, F)$ be a DFA and $L = L(A)$. Then it is clear that if $\delta(0, x) = \delta(0, y) = q$ for some $q \in Q$, then $x \equiv_L y$ and, thus, $x \sim_L y$. Therefore, we can also define equivalence as well as similarity relations on states.

Definition 2. *Let $A = (Q, \Sigma, \delta, 0, F)$ be a DFA. We define, for each state $q \in Q$,*

$$level(q) = min\{|w| \mid \delta(0, w) = q\},$$

i.e., $level(q)$ is the length of the shortest path from the initial state to q.

Definition 3. *Let $A = (Q, \Sigma, \delta, 0, F)$ be a DFA and $L = L(A)$. We say that $p \equiv_A q$ (state p is equivalent to q in A) if for every $w \in \Sigma^*$, $\delta(s, w) \in F$ iff $\delta(q, w) \in F$.*

Definition 4. *Let $A = (Q, \Sigma, \delta, 0, F)$ be a DFCA of a finite language L. Let $level(p) = i$ and $level(q) = j$, $m = max\{i, j\}$. We say that $p \sim_A q$ (state p is L-similar to q in A) if for every $w \in \Sigma^{\leq l-m}$, $\delta(p, w) \in F$ iff $\delta(q, w) \in F$.*

If $A = (Q, \Sigma, \delta, 0, F)$ is a DFA, for each $q \in Q$, we denote $x_A(q) = min\{w \mid \delta(0, w) = q\}$, where the minimum is taken according to the quasi-lexicographical order, and $L_A(q) = \{w \in \Sigma^* \mid \delta(q, w) \in F\}$. When the automaton A is understood, we write x_q instead of $x_A(q)$ and L_q instead $L_A(q)$.

Lemma 2 *Let $A = (Q, \Sigma, \delta, 0, F)$ be a DFCA of a finite language L. Let $x, y \in \Sigma^*$ such that $\delta(0, x) = p$ and $\delta(0, y) = q$. If $p \sim_A q$ then $x \sim_L y$.*

Proof. Let $level(p) = i$ and $level(q) = j$, $m = max\{i, j\}$, and $p \sim_A q$. Choose an arbitrary $w \in \Sigma^*$ such that $|xw| \leq l$ and $|yw| \leq l$. Because $i \leq |x|$ and $j \leq |y|$ it follows that $|w| \leq l - m$. Since $p \sim_A q$ we have that $\delta(p, w) \in F$ iff $\delta(q, w) \in F$, i.e. $\delta(0, xw) \in F$ iff $\delta(0, yw) \in F$, which means that $xw \in L(A)$ iff $yw \in L(A)$. Hence $x \sim_L y$.

Lemma 3 *Let $A = (Q, \Sigma, \delta, 0, F)$ be DFCA of a finite language L. Let $level(p) = i$ and $level(q) = j$, $m = max\{i, j\}$, and $x \in \Sigma^i$, $y \in \Sigma^j$ such that $\delta(0, x) = p$ and $\delta(0, y) = q$. If $x \sim_L y$ then $p \sim_A q$.*

Proof. Let $x \sim_L y$ and $w \in \Sigma^{\le l-m}$. If $\delta(p, w) \in F$, then $\delta(0, xw) \in F$. Because $x \sim_L y$, it follows that $\delta(0, yw) \in F$, so $\delta(q, w) \in F$. Using the symmetry we get that $p \sim_A q$. ¿

Corollary 2. *Let* $A = (Q, \Sigma, \delta, 0, F)$ *be a DFCA of a finite language* L*. Let* $level(p) = i$ *and* $level(q) = j$*,* $m = \max\{i, j\}$*, and* $x_1 \in \Sigma^i$*,* $y_1 \in \Sigma^j$*,* $x_2, y_2 \in \Sigma^*$*, such that* $\delta(0, x_1) = \delta(0, x_2) = p$ *and* $\delta(0, y_1) = \delta(0, y_2) = q$*. If* $x_1 \sim_L y_1$ *then* $x_2 \sim_L y_2$*.*

Example 2. If x_1 and y_1 are not minimal, i.e. $|x_1| > i$, but $p = \delta(0, x_1)$ or $|y_1| > j$, but $q = \delta(0, y_1)$, then the conclusion of Corollary 2 is not true.

Let $L = \{a, b, aa, aaa, bab\}$, so $l = 3$ (Figure 1).

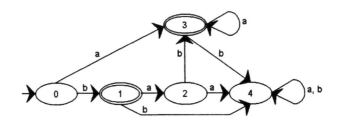

Fig. 1. A DFCA of L

we have that $b \sim_L bab$, but $b \not\sim_L a$.

Corollary 3. *Let* $A = (Q, \Sigma, \delta, 0, F)$ *be a DFCA of a finite language* L *and* $p, q \in Q$*,* $p \neq q$*. Then* $x_p \sim_L x_q$ *iff* $p \sim_A q$*.*

If $p \sim_A q$, and $level(p) \le level(q)$ and $q \in F$ then $p \in F$.

Lemma 4 *Let* $A = (Q, \Sigma, \delta, 0, F)$ *be a DFCA of a finite language* L*. Let* $s, p, q \in Q$ *such that* $level(s) = i$*,* $level(p) = j$*,* $level(q) = k$*,* $i \le j \le k$*. The following statements are true:*

1. *If* $s \sim_A p$*,* $s \sim_A q$*, then* $p \sim_A q$*.*
2. *If* $s \sim_A p$*,* $p \sim_A q$*, then* $s \sim_A q$*.*
3. *If* $s \sim_A p$*,* $p \not\sim_A q$*, then* $s \not\sim_A q$*.*

Proof. We apply Lemma 1 and Corollary 3.

Lemma 5 *Let* $A = (Q, \Sigma, \delta, 0, F)$ *be a DFCA of a finite language* L*. Let* $level(p) = i$*,* $level(q) = j$*, and* $m = \max\{i, j\}$*. If* $p \sim_A q$ *then* $L_p \cap \Sigma^{\le l-m} = L_q \cap \Sigma^{\le l-m}$ *and* $L_p \cup L_q$ *is a L- similarity set.*

The proof is left to the reader.The next lemma is obvious.

Lemma 6 *Let $A = (Q, \Sigma, \delta, 0, F)$ be a DFCA of a finite language L. Let $i = level(p)$ and $j = level(q)$, $i \leq j$. Let $p \sim_L q$.Let $w = w_1 \ldots w_n \in \Sigma^{\leq l}$ and $p_i = \delta(0, w_1 \ldots w_i)$, $1 \leq i \leq n$. Then $w \in L$ iff $x_k w_{k+1} \ldots w_n \in L$ for $1 \leq k \leq n$.*

Lemma 7 *Let $A = (Q, \Sigma, \delta, 0, F)$ be a DFCA of a finite language L. If $p \sim_A q$ for some $p, q \in Q$, $i = level(p)$, $j = level(q)$ and $i \leq j$, $p \neq q$, $q \neq 0$. then we can construct a DFCA $A' = (Q', \Sigma, \delta', 0, F')$ of L such that $Q' = Q - \{q\}$, $F' = F - \{q\}$, and*

$$\delta'(s, a) = \begin{cases} \delta(s, a) & \text{if } \delta(s, a) \neq q, \\ p & \delta(s, a) = q \end{cases}$$

for each $s \in Q'$ and $a \in \Sigma$. Thus, A is not a minimal DFCA of L.

Proof. It suffices to prove that A' is a DFCA of L. Let l be the length of the longest word(s) in L and assume that $level(p) = i$ and $level(q) = j$, $i \leq j$. Consider a word $w \in \Sigma^{\leq L}$. We now prove that $w \in L$ iff $\delta'(0, w) \in F'$.

If there is no prefix w_1 of w such that $\delta(0, w_1) = q$, then clearly $\delta'(0, w) \in F'$ iff $\delta(0, w) \in F$. Otherwise, let $w = w_1 w_2$ where w_1 is the shortest prefix of w such that $\delta(0, w_1) = q$. In the remaining, it suffices to prove that $\delta'(p, w_2) \in F'$ iff $\delta(q, w_2) \in F$. We prove this by induction on the length of w_2. First consider the case $|w_2| = 0$, i.e., $w_2 = \lambda$. Since $p \sim_A q$, $p \in F$ iff $q \in F$. Then $p \in F'$ iff $q \in F$ by the construction of A'. Thus, $\delta'(p, w_2) \in F'$ iff $\delta(q, w_2) \in F$. Suppose that the statement holds for $|w_2| < l'$ for $l' \leq l - |w_1|$. (Note that $l - |w_1| \leq l - j$.) Consider the case that $|w_2| = l'$. If there does not exist $u \in \Sigma^+$ such that $u \preceq_p w_2$ and $\delta(p, u) = q$, then $\delta(p, w_2) \in F - \{q\}$ iff $\delta(q, w_2) \in F - \{q\}$, i.e., $\delta'(p, w_2) \in F'$ iff $\delta(q, w_2) \in F$. Otherwise, let $w_2 = uv$ and u be the shortest nonempty prefix of w_2 such that $\delta(p, u) = q$. Then $|v| < l'$ (and $\delta'(p, u) = p$). By induction hypothesis, $\delta'(p, v) \in F'$ iff $\delta(q, v) \in F$. Therefore, $\delta'(p, uv) \in F'$ iff $\delta(q, uv) \in F$.

Lemma 8 *Let A be a DFCA of L and $L' = L(A)$. Then $x \equiv_{L'} y$ implies $x \sim_L y$.*

Proof. It is clear that if $x \equiv_L y$ then $x \sim_L y$. Let l be the length of the longest word(s) in L. Let $x \equiv_{L'} y$. So, for each $z \in \Sigma^*$, $xz \in L'$ iff $yz \in L'$. We now consider all words $z \in \Sigma^*$, such that $| xz | \leq l$ and $| yz | \leq l$. Since $L = L' \cap \Sigma^{\leq l}$ and $xz \in L'$ iff $yz \in L'$, we have $xz \in L$ iff $yz \in L$. Therefore, $x \sim_L y$ by the definition of $\sim_L$.

Corollary 4. *Let $A = (Q, \Sigma, \delta, 0, F)$ be a DFCA of a finite language L, $L' = L(A)$. Then $p \equiv_A q$ implies $p \sim_L q$.*

Corollary 5. *A minimal DFCA of L is a minimal DFA.*

Proof. Let $A = (Q, \Sigma, \delta, 0, F)$ be a minimal DFCA of a finite language L. Suppose that A is not minimal as a DFA for $L(A)$, then there exists $p, q \in Q$ such that $p \equiv_{L'} q$, then $p \sim_L q$. By Lemma 7 it follows that A is not a minimal DFCA, contradiction.

Remark 1. Let A be a DFCA of L and A a minimal DFA. Then A may not be a minimal DFCA of L.

Example 3. We take the DFA's of Figure 2.

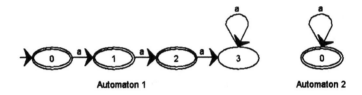

Automaton 1   Automaton 2

Fig. 2. Example

The DFA in Automaton 1 is a minimal DFA and a DFCA of $L = \{\lambda, a, aa\}$ but not a minimal DFCA of L, since the DFA in Automaton 2 is a minimal DFCA of L:

Theorem 3. *Any minimal DFCA of L has exactly $N(L)$ states.*

Proof. Let $A = (Q, \Sigma, \delta, 0, F)$ be DFCA of a finite language L, and $\#Q = n$.

Suppose that $n > N(L)$. Then there exist $p, q \in Q$, $p \neq q$, such that $x_p \sim_L x_q$ (because of the definition of $N(L)$). Then $p \sim_A q$ by Lemma 3. Thus, A is not minimal. A contradiction.

Suppose that $N(L) > n$. Let $[y_1, \ldots, y_{N(L)}]$ be a canonical dissimilar sequence of L. Then there exist i, j, $1 \leq i, j \leq N(L)$ and $i \neq j$, such that $\delta(0, y_i) = \delta(0, y_j) = q$ for some $q \in Q$. Then $y_i \sim_L y_j$. Again a contradiction.

Therefore, we have $n = N(L)$.

5   The Construction of Minimal DFCA

The first part of this section describe an algorithm that determines the similarity relations between states. The second part is to construct a minimal DFCA assuming that the similarity relation between states is known.

An ordered DFA is a DFA where $\delta(i, a) = j$ implies that $i \leq j$, for all states i, j and letters a.

5.1 Determining Similarity Relation between States

The aim is to present an algorithm which determines the similarity relations between states.

Let $A = (\Sigma, Q, 0, \delta, F)$ a DFCA of a finite language L. For each $s \in Q$ let $\gamma_s = \min\{w \mid \delta(s, w) \in F\}$, where minimum is taken according to the quasi-lexicographical order. Define $D_i = \{s \in Q \mid |\gamma_s| = i\}$, for each $i = 0, 1, \dots$.

Lemma 9 *Let $A = (\Sigma, Q, 0, \delta, F)$ a DFCA of a finite language L, and $s \in D_i$, $p \in D_j$. If $i \neq j$ then $s \not\sim p$.*

Proof. We can assume that $i < j$. Then obviously $\delta(s, \gamma_s) \in F$ and $\delta(p, \gamma_s) \notin F$. Since $l \geq |x_s| + \gamma_s|$, $l \geq |x_p| + |\gamma_p|$, and $i < j$, it follows that $|\gamma_s| < |\gamma_p|$. So, we have that $|\gamma_s| \leq \min(l - |x_s|, l - |x_p|)$. Hence, $s \not\sim p$.

Lemma 10 *Let $A = (Q, \Sigma, 0, \delta, F)$ be a reduced ordered DFA accepting L, $p, q \in Q - \{\#Q - 1\}$, where $\#Q - 1$ is the sink state, and either $p, q \in F$ or $p, q \notin F$. If for all $a \in \Sigma$, $\delta(p, a) \sim_A \delta(q, a)$, then $p \sim_A q$.*

Proof. Let $a \in \Sigma$ and $\delta(p, a) = r$ and $\delta(q, a) = s$. If $r \sim_A s$ then for all $|w|$, $|w| < l - \max\{x_A(s), x_A(r)\}$, $x_A(r)w \in L$ iff $x_A(s)w \in L$. Using Lemma 2 we also have: $x_A(q)aw \in L$ iff $x_A(s)w \in L$ for all $w \in \Sigma^*$, $|w| \leq l - |x_A(s)|$ and $x_A(p)aw \in L$ iff $x_A(r)w \in L$ for all $w \in \Sigma^*$, $|w| \leq l - |x_A(r)|$.

Hence $x_A(p)aw \in L$ iff $x_A(q)aw \in L$, for all $w \in \Sigma^*$, $|w| \leq l - \max\{|x_A(r)|, |x_A(s)|\}$. Because $|x_A(r)| \leq |x_A(q)a| = |x_A(q)| + 1$ and $|x_A(s)| \leq |x_A(p)a| = |x_A(p)| + 1$, we get $x_A(p)aw \in L$ iff $x_A(q)aw \in L$, for all $w \in \Sigma^*$, $|w| \leq l - \max\{|x_A(p)|, |x_A(q)|\} - 1$.

Since $a \in \Sigma$ is chosen arbitrary, we conclude that $x_A(p)w \in L$ iff $x_A(q)w \in L$, for all $w \in \Sigma^*$, $|w| \leq l - \max\{|x_A(p)|, |x_A(q)|\}$, i.e. $x_A(p) \sim_A x_A(q)$. Therefore, by using Lemma 3, we get that $p \sim_A q$.

Lemma 11 *Let $A = (Q, \Sigma, 0, \delta, F)$ be a reduced ordered DFA accepting L such that $\delta(0, w) = s$ implies $|w| = |x_s|$ for all $s \in Q$. Let $p, q \in Q - \{\#Q - 1\}$, where $\#Q - 1$ is the sink state. If there exists $a \in \Sigma$ such that $\delta(p, a) \not\sim_A \delta(q, a)$, then $p \not\sim_A q$.*

Proof. Suppose that $p \sim_A q$. then for all $aw \in \Sigma^{l-m}$, $\delta(p, aw) \in F$ iff $\delta(q, aw) \in F$, where $m = \max\{level(p), level(q)\}$. So $\delta(\delta(p, a), w) \in F$ iff $\delta(\delta(q, a), w) \in F$ for all $w \in \Sigma^{l-m-1}$. Since $|x_{\delta(p,a)}| = |x_p| + 1$ and $|x_{\delta(q,a)}| = |x_q| + 1$ it follows by definition that $\delta(p, a) \sim_A \delta(q, a)$. This is a contradiction.

Our algorithm for determining the similarity relation between the states of a DFA (DFCA) of a finite language is based on Lemmas 10 and 11. However, most of DFA (DFCA) do not satisfy the condition of Lemma 11. So, we shall first transform the given DFA (DFCA) into one that does.

Let $A = (Q_A, \Sigma, 0, \delta_A, F_A)$ be a DFCA of L. We construct the minimal DFA for the language $\Sigma^{\leq l}$, $B = (Q_B, \Sigma, 0, \delta_B, F_B)$ ($Q_B = \{0, \dots, l, l + 1\}$,

$\delta_B(i, a) = i + 1$, for all i, $0 \leq i \leq l$, $\delta_B(l + 1, a) = l + 1$, for all $a \in \Sigma$, $F_B = \{0, \ldots, l\}$). The DFA B will have exact $l + 2$ states.

Now we use the standard Cartesian product construction (see, e.g., [3], for details) for the DFA $C = (Q_C, \Sigma, q_0, \delta_C, F_C)$ such that $L(C) = L(A) \cap L(B)$, and we eliminate all inaccessible states. Obviously, $L(C) = L$ and C satisfies the condition of Lemma 11.

The next lemma is easy to prove and left for the reader.

Lemma 12 *For the DFA C constructed above we have $(p, q) \sim_C (p, r)$.*

Lemma 13 *For the DFA C constructed above, if $\delta_C((0, 0), w) = (p, q)$, then $|w| = q$.*

Proof. We have $\delta_C((0, 0), w) = (p, q)$, so $\delta_B(0, w) = q$ so $|w| = q$.

Now we are able to present an algorithm, which determines the similarity relation between the states of C. Note that Q_C is ordered by that $(p_A, p_B) < (q_A, q_B)$ if $p_A < q_A$ or $p_A = q_A$ and $p_B < q_B$. Attaching to each state of C is a list of similar states. For $\alpha, \beta \in Q_C$, if $\alpha \sim_C \beta$ and $\alpha < \beta$, then β is stored on the list of similar states for α.

We assume that $Q_A = \{0, 1, \ldots, n\}$ and n is the sink state of A.

1. Generate the DFA B for the language $\Sigma^{\leq l}$.
2. Compute the DFA C such that $L(C) = L(A) \cap L(B)$ using the standard Cartesian product algorithm (see [3] for details).
3. Compute D_i of C, $0 \leq i \leq l$.
4. Initialize the similarity relation by specifying:
   - For all $(n, p), (n, q) \in Q_C$, $(n, p) \sim_C (n, q)$.
   - For all $(n, l + 1 - i) \in Q_C$, $(n, l + 1 - i) \sim_C \alpha$ for all $\alpha \in D_j$, $j = i, \ldots, l$, $0 \leq i \leq l$.
5. For each D_i, $0 \leq i \leq l$, create a list $List_i$, which is initialized to $\emptyset$.
6. For each $\alpha \in Q_C - \{(n, q) \mid q \in Q_B\}$, following the reversed order of Q_C, do the following: Assuming $\alpha \in D_i$.
   - For each $\beta \in List_i$, if $\delta_C(\alpha, a) \sim_C \delta_C(\beta, a)$ for all $a \in \Sigma$, then $\alpha \sim_C \beta$.
   - Put α on the list $List_i$.

Remark 2. The above algorithm has complexity $O((n \times l)^2)$, where n is the number of states of the initial DFA (DFCA) and l is the maximum accepted length for the finite language L.

5.2   The Construction of a Minimal DFCA

As input we have the above DFA C and, with each $\alpha \in Q_C$, a set $S_\alpha = \{\beta \in Q_C \mid \alpha \sim_C \beta$ and $\alpha < \beta\}$. The output is $D = (Q_D, \Sigma, \delta_D, q_0, F_D)$, a minimal DFCA for L.

We define the following:

$i = 0, q_i = 0, T = Q_C - S_i, (x_0 = \lambda)$;

while $(T \neq \emptyset)$ do the following:

$\quad i = i + 1$;

$\quad q_i = \min\{s \in T\}$,

$\quad T = T - S_{q_i}, (x_i = \min\{w \mid \delta_C(0, w) \in S_i\})$;

$m = i$;

Then $Q_D = \{q_0, \ldots, q_m\}$; $q_0 = 0$; $\delta_D(i, a) = j$, iff $k = \min S_i$ and $\delta_C(k, a) \in S_j$; $F_D = \{i \mid S_i \cap F_C \neq \emptyset\}$.

Note that the constructions of x_i above are useful for the proofs in the following only, where the min (minimum) operator for x_i is taken according to the lexicographical order. Let $X_i = \{(i, s) \mid (i, s) \in Q_C\}$ and $a_i = \#X_i, 0 \leq i \leq l+1$.

Step 1. For all $1 \leq i \leq l + 1$ do $b_i = a_i$, for all $(i, j) \in Q_C$ do $new((i, j)) = -1$. Set $m = 0$, $r = 0$ and $s = 0$.

Step 2. Put $S_m = \{(p, q) \in Q_C \mid (r, s) \sim_A (p, q)\}$.

Step 3. For all $(p, q) \in S_m$, perform $new((p, q)) = m$ and $b_p = b_p - 1$.

Step 4. Put $m = m + 1$.

Step 5. While $b_r = 0$ and $r \leq l+1$ do $r = r + 1$. If $r > l + 1$ then go to Step 7, else go to Step 6.

Step 6. Take the state $(r, s) \in A_r$ such that $new(r, s) \neq -1$, and s is the minimal with this property. Go to Step 2.

Step 7. $Q_D = \{0, \ldots, m - 1\}$, $F = \{i \mid new((p, q)) = i, (p, q) \in F_C\}$. For all $q \in Q_D$ and $a \in \Sigma$ set $\delta_D(q, a) = -1$.

Step 8. For all $p = 0, \ldots, l + 1$, $q = 0, \ldots, n$, $(p, q) \in Q_C$ and $a \in \Sigma$if $\delta_D(new(p, q), a) = -1$ define

$$\delta_D(new(p, q), a) = new(\delta_C((p, q), a)).$$

According to the algorithm we have a total ordering of the states Q_C: $(p, q) \leq (r, s)$ if $(p, q) = (r, s)$ or $p < r$ or $p = r$ and $q < s$. Hence $\delta_D(i, a) = j$ iff $\delta_D(0, x_i a) = j$. Also, using the construction (i.e. the total order on Q_C) it follows that $0 = |x_0| \leq |x_1| \leq \ldots \leq |x_{m-1}|$.

Lemma 14 *The sequence* $[x_0, x_1, \ldots, x_{m-1}]$, *constructed above is a cannonical L- dissimilar sequence.*

Proof. We construct the sets $X_i = \{w \in \Sigma^* \mid \delta(0, w) \in S_i\}$. Obviously $X_i \neq \emptyset$. From Lemma 2 it follows that X_i is a L- similarity set for all $0 \leq i \leq m - 1$.

Let $w \in \Sigma^*$. Because $(S_i)_{1 \leq i \leq m-1}$ is a partition of Q, $w \in X_i$ for some $0 \leq i \leq n - 1$, so $(X_i)_{0 \leq i \leq n-1}$ is a partition of Σ^* and therefore a cannonical L-dissimilar sequence.

Corollary 6. *The automaton D constructed above is a minimal DFCA for L.*

Proof. Since the number of states is equal to the number of elements of a cannonical L-dissimilar sequence, we only have to prove that D is a cover automaton for L. Let $w \in \Sigma^{\leq l}$. We have that $\delta_D(0, w) \in F_D$ iff $\delta_C((0, 0), w) \in S_f$ and $S_f \cap F_C \neq \emptyset$, i.e. $x_f \sim_C w$. Since $|w| \leq l$, $x_f \in L$ iff $w \in L$ (because C is a DFCA for L).

6 Boolean Operations

We shall use similar constructions as in [3] for constructing DFCA of languages which are a result of boolean operations between finite languages. The modifications are suggested by the previous algorithm. We first construct the DFCA which satisfies hypothesis of Lemma 11 and afterwards we can minimise it using the general algorithm. Since the minimisation will follow in a natural way we shall present only the construction of the necessarily DFCA.

Let $A_i = (Q_i, \Sigma, 0, \delta_i, F_i)$, two DFCA of the finite languages L_i, $l_i = \max\{|w| \mid w \in L_i\}$, $i = 1, 2$.

6.1 Intersection

We construct the following DFA:

$A = (Q_1 \times Q_2 \times \{0, \ldots, l\}, \Sigma, \delta, (0, 0, 0), F)$, where
$l = \min\{l_1, l_2\}$, $\delta((s, p, q), a) = (\delta_1(s, a), \delta_2(p, a), q + 1)$, for $s \in Q_1$, $p \in Q_2$, $q \leq l$, and $\delta((s, p, l + 1), a) = (\delta_1(s, a), \delta_2(p, a), l + 1)$ and $F = \{(s, p, q) \mid s \in F_1, p \in F_2, q \leq l\}$.

Theorem 4. *The automaton A constructed above is a DFA for $L = L(A_1) \cap L(A_2)$.*

Proof. We have the following relations: $w \in L_1 \cap L_2$ iff $|w| \leq l$ and $w \in L_1$ and $w \in L_2$ iff $|w| \leq l$ and $w \in L(A_1)$ and $w \in L(A_2)$. The rest of the proof is obvious.

6.2 Union

We construct the following DFA:

$A = (Q_1 \times Q_2 \times \{0, \ldots, l\}, \Sigma, \delta, (0, 0, 0), F)$, where
$l = \max\{l_1, l_2\}$, $m = \min\{l_1, l_2\}$, $\delta((s, p, q), a) = (\delta_1(s, a), \delta_2(p, a), q + 1)$, for $s \in Q_1$, $p \in Q_2$, $q \leq l$, and $\delta((s, p, l + 1), a) = (\delta_1(s, a), \delta_2(p, a), l + 1)$ and $F = \{(s, p, q) \mid s \in F_1 \text{ or } p \in F_2, \ q \leq m\} \cup \{(s, p, q) \mid s \in F_r \text{ and } m < q \leq l\}$, where r is such that $l_r = l$.

Theorem 5. *The automaton A constructed above is a DFA for $L = L(A_1) \cup L(A_2)$.*

Proof. We have the following relations: $w \in L_1 \cup L_2$ iff $|w| \leq m$ and $w \in L_1$ or $w \in L_2$, or $m < |w| \leq l$ and $w \in L_r$ iff $|w| \leq m$ and $w \in L(A_1)$ or $w \in L(A_2)$, or $m < |w| \leq l$ and $w \in L(A_r)$. The rest of the proof is obvious.

6.3 Symmetric Difference

We construct the following DFA:

$A = (Q_1 \times Q_2 \times \{0, \ldots, l\}, \Sigma, \delta, (0, 0, 0), F)$, where
$l = \max\{l_1, l_2\}$, $m = \min\{l_1, l_2\}$, $\delta((s, p, q), a) = (\delta_1(s, a), \delta_2(p, a), q + 1)$, for $s \in Q_1$, $p \in Q_2$, $q \leq l$, and $\delta((s, p, l + 1), a) = (\delta_1(s, a), \delta_2(p, a), l + 1)$ and $F = \{(s, p, q) \mid s \in F_1 \text{ or exclusive } p \in F_2, \ q \leq m\} \cup \{(s, p, q) \mid s \in F_r \text{ and } m < q \leq l\}$, where r is such that $l_r = l$.

Theorem 6. *The automaton A constructed above is a DFA for $L = L(A_1)\Delta$ $L(A_2)$.*

Proof. We have the following relations: $w \in L_1 \Delta L_2$ iff $|w| \leq m$ and $w \in L_1$ or exclusive $w \in L_2$, or $m < |w| \leq l$ and $w \in L_r$ iff $|w| \leq m$ and $w \in L(A_1)$ or exclusive $w \in L(A_2)$, or $m < |w| \leq l$ and $w \in L(A_r)$. The rest of the proof is obvious.

6.4 Difference

We construct the following DFA:

$A = (Q_1 \times Q_2 \times \{0, \dots, l\}, \Sigma, \delta, (0,0,0), F)$, where

$l = \max\{l_1, l_2\}$, $m = \min\{l_1, l_2\}$ and $\delta((s,p,q), a) = (\delta_1(s,a), \delta_2(p,a), q+1)$, for $s \in Q_1$, $p \in Q_2$, $q \leq l$, and $\delta((s,p,l+1), a) = (\delta_1(s,a), \delta_2(p,a), l+1)$. If $l_1 < l_2$ then $F = \{(s,p,q) \mid s \in F_1 \text{ and } p \notin F_2, q \leq m\}$ and $F = \{(s,p,q) \mid s \in F_1 \text{ and } p \notin F_2, q \leq m\} \cup \{(s,p,q) \mid s \in F_1 \text{ and } m < q \leq l\}$, if $l_1 \geq l_2$.

Theorem 7. *The automaton A constructed above is a DFA for $L = L(A_1) - L(A_2)$.*

Proof. We have the following relations: $w \in L_1 - L_2$ iff $|w| \leq m$ and $w \in L_1$ and $w \notin L_2$, or $m < |w| \leq l$ and $w \in L_1$ iff $|w| \leq m$ and $w \in L(A_1)$ and $w \notin L(A_2)$, or $m < |w| \leq l$ and $w \in L(A_1)$. The rest of the proof is obvious.

Open Problems 1) Try to find a better algorithm for minimisation 2) or prove that any minimisation algorithm has complexity $\Omega(n^2)$. 3) Find a better algorithm for determining similar states 4) in any DFCA of L. 3) Find better algorithms for boolean operations on DFCA.

References

1. J.L. Balcàzar, J. Diaz, and J. Gabarrò, Uniform characterisations of non-uniform complexity measures, *Information and Control* 67 (1985) 53-89.
2. Y. Breitbart, On automaton and "zone" complexity of the predicate "tobe a kth power of an integer",*Dokl. Akad. Nauk SSSR***196** (1971), 16-19[Russian]; Engl. transl.,Soviet Math. Dokl. **12** (1971), 10-14.
3. Cezar Câmpeanu. Regular languages and programming languages, *Revue Roumaine de Linguistique - CLTA*, 23 (1986), 7-10.
4. C.Dwork and L.Stockmeyer, A time complexity gap for two-way probabilistic finite-state automata, *SIAM Journal on Computing* 19 (1990) 1011-1023.
5. J. Hartmanis, H.Shank, Two memory bounds for the recognition of primes by automata, *Math. Systems Theory* **3** (1969), 125-129.
6. J.E. Hopcroft and J.D. Ullman, Introduction to Automata Theory, Languages, and Computation, Addison Wesley (1979), Reading, Mass.
7. J. Kaneps, R. Freivalds, Minimal Nontrivial Space Space Complexity of Probabilistic One-Way Turing Machines, in Proceedings of Mathematical Foundations of Computer Science, Banská Bystryca, Czechoslovakia, August 1990, *Lecture Notes in Computer Science*, vol 452, pp. 355-361, Springer-Verlag, New York/Berlin, 1990.

8. J. Kaneps, R. Freivalds, Running time to recognise non-regular languages by 2-way probabilistic automata, in ICALP'91, *Lecture Notes in Computer Science*, vol 510, pp. 174-185, Springer-Verlag, New York/Berlin, 1991.

9. J. Paredaens, R. Vyncke, A class of measures on formal languages, *Acta Informatica*, **9** (1977), 73-86.

10. Jeffrey Shallit, Yuri Breitbart, Automaticity I: Properties of a Measure of Descriptional Complexity, *Journal of Computer and System Sciences*, **53**, 10-25 (1996).

11. A. Salomaa, *Theory of Automata*, Pergamon Press (1969), Oxford.

12. K. Salomaa, S. Yu, Q. Zhuang, The state complexities of some basic operations on regular languages, *Theoretical Computer Science* 125 (1994) 315-328.

13. B.A. Trakhtenbrot, Ya. M. Barzdin, Finite Automata: Behaviour and Synthesis, *Fundamental Studies in Computer Science*, Vol.1, North-Holland, Amsterdam, 1973.

14. S. Yu, Q. Zhung, On the State Complexity of Intersection of Regular Languages, *ACM SIGACT News*, vol. 22, no. 3, (1991) 52-54.

15. S. Yu, Regular Languages, Handbook of Formal Languages, Springer Verlag, 1995.

Determinization of Glushkov Automata[*]

Jean-Marc Champarnaud, Djelloul Ziadi, and Jean-Luc Ponty

LIFAR
Université de Rouen, Faculté des Sciences et des Techniques
76821 Mont-Saint-Aignan Cedex, France
{champarnaud, ponty, ziadi}@dir.univ-rouen.fr

Abstract. We establish a new upper bound on the number of states of the automaton yielded by the determinization of a Glushkov automaton. We show that the ZPC structure, which is an implicit construction for Glushkov automata, leads to an efficient implementation of the subset construction.

1 Introduction

Automata determinization may be exponential, whereas most of automata operations are polynomial. It is not possible to avoid this behaviour in the general case [16]. Therefore it is quite natural to take great care of the implementation of determinization algorithm, so that this transformation penalizes as little as possible the performances of an automata software.

Two different kinds of computations are carried along the determinization process: the computation of the transition of a state of the deterministic automaton by a letter (which yields a subset of states of the nondeterministic automaton), and the set equality tests which make it possible to decide whether a transition generates a new state or not.

Concerning the computation of the transitions, the choice of the data structure used to implement the transitions of the nondeterministic automaton has a lot of influence on the performances. The AUTOMATE [4], INR [8] and Grail [17] softwares make use of a representation in which both transitions with the same origin and transitions with the same origin and the same letter are contiguous. Jonhson and Wood [9] have studied the efficiency of sorting procedures applied to subsets computed from such data structures.

As for the set equality tests, the choice of the data structure for memorizing the subsets is obviously an important factor of the complexity. According to Ponty [14], the number of integer comparisons involved by set equality tests is $O(\sqrt{n}2^{2n})$ if arrays are used and $O(n^2(\log n)2^n)$ if binary search trees are used. Such complexities justify the work of Leslie [11] and Leslie, Raymond and Wood [12].

---

[*] This work is a contribution to the Automate software development project carried on by A.I.A. Working Group (Algorithmics and Implementation of Automata), L.I.F.A.R. Contact: {Champarnaud, Ziadi}@dir.univ-rouen.fr.

J.-M. Champarnaud, D. Maurel, D. Ziadi (Eds.): WIA'98, LNCS 1660, pp. 57–68, 1999.

In this paper we consider a particular family of automata, which are computed from regular expressions following the Glushkov algorithm. Glushkov automata have been characterized by Caron and Ziadi [3] in terms of graphs. On the other hand, Ziadi *et al.* [15], have designed a time and space linear representation of the Glushkov automaton of a regular expression, which is based on two forests of states and a set of links going from one forest to the other one. The ZPC structure leads to an output sensitive implementation of the conversion of a regular expression into an automaton [15] and to an efficient algorithm for testing the membership of a word to a regular language [14].

We shall show first that the number of states of the automaton resulting from the determinization of a Glushkov automaton can be bounded more tightly than in the general case. Then we shall make use of the properties of the ZPC structure to improve the complexity of the computation of the transitions, as well as the complexity of set equality tests.

The next section recalls some useful definitions and notations, as well as the Glushkov algorithm and the ZPC structure design. Section 3 describes the subset construction. Section 4 gathers our results about Glushkov automata determinization.

2   Definitions and Notations

We shall limit ourselves to definitions involved by the description of the new algorithms. For further details about regular languages and finite automata, the references [7,18,5,13] should be consulted.

A finite automaton is a 5-tuple $\mathcal{M} = (Q, \Sigma, \delta, I, F)$ where Q is a (finite) set of states, Σ is a (finite) alphabet, $I \subseteq Q$ is the set of initial states, $F \subseteq Q$ is the set of terminal states, and δ is the transition relation. A deterministic finite automaton (DFA) has a unique initial state and arrives in a unique state (if any) after scanning a symbol of Σ. Otherwise the automaton is nondeterministic (NFA). The language $L(\mathcal{M})$ recognized by the automaton $\mathcal{M}$ is the set of words of Σ^* whose scanning makes $\mathcal{M}$ arrive to a terminal state.

A regular expression over an alphabet Σ is generated by recursively applying operators '+' (union), '·' (concatenation) and '*' (Kleene star) to atomic expressions (every symbol of Σ, the empty word and the empty set). A language is regular if and only if it can be denoted by a regular expression. The *length* of a regular expression E, denoted $|E|$, is the number of operators and symbols in E. The *alphabetic width* of E, denoted $||E||$, is the number of symbols in E.

The Kleene theorem [10] states that a language is regular if and only if it is recognized by a finite automaton. Computing the Glushkov automaton of a regular expression [6] is a constructive proof of this theorem.

2.1   Glushkov Automaton

Glushkov algorithm works on a linearized expression E' deduced from E by ranking every symbol occurrence with its position in E. For example: if $E = a(b + a)^* + a$ then $E' = a_1(b_2 + a_3)^* + a_4$.

The set of positions of E is denoted $Pos(E)$. The application $\chi : Pos(E) \to \Sigma$ maps every position to its *value* in Σ.

Algorithm Glushkov(E)

1. Linearize the expression E. The result is the expression E'.
2. Compute the following sets:
   - $Null_E$ which is $\{\varepsilon\}$ if $\varepsilon \in L(E)$ and $\emptyset$ otherwise.
   - $First(E)$, the set of positions that match the first symbol of some word in $L(E')$.
   - $Last(E)$, the set of positions that match the last symbol of some word in $L(E')$.
   - $Follow(E, x)$, $\forall x \in Pos(E)$: the set of positions that follow the position x in some word of $L(E')$.
3. Compute the Glushkov automaton of E, $\mathcal{M}_E = (Q, \Sigma, \delta, s_I, F)$ where:
   - $Q = Pos(E) \cup \{s_I\}$
   - $\forall a \in \Sigma, \delta(s_I, a) = \{x \in First(E) \mid \chi(x) = a\}$
   - $\forall x \in Q, \forall a \in \Sigma, \delta(x, a) = \{y \mid y \in Follow(E, x) \text{ and } \chi(y) = a\}$
   - $F = Last(E) \cup Null_E \cdot \{s_I\}$

2.2   The ZPC Structure

The ZPC structure of a regular expression E is based on two forests deduced from its syntax tree $T(E)$. These forests respectively encode the Last sets and the First sets associated to the subexpressions of E. The transition relation of the Glushkov automaton of E naturally appears as a collection of links from the Last forest to the First forest.

Let us sketch the construction of the ZPC structure.

Algorithm ZPC(E)

1. Compute the syntax tree $T(E)$.
2. Compute the forests $TL(E)$ and $TF(E)$.
3. Compute the set of follow links going from $TL(E)$ to $TF(E)$.
4. Remove redundant follow links.

The Lasts forest $TL(E)$ is a copy of $T(E)$, where a link going from a node labeled '·' to its left child is deleted if the language of its right child does not contain ε. Thus the property: $Last((F) \cdot (G)) = Last(G) \cup Null_G \cdot Last(F)$ is satisfied. Furthermore, each node of $TL(E)$ points to its leftmost and rightmost leaves, and leaves in the same Last set are linked.

The Firsts forest $TF(E)$ is computed in a similar way, by deleting a link going from a node labeled '·' to its right child, if the language of its left child does not contain ε, w.r.t. the property: $First((F) \cdot (G)) = First(F) \cup Null_F \cdot First(G)$.

The two forests are connected as follows. If a node of $TL(E)$ is labeled by '·', its left child is linked to the right child of the corresponding node in $TF(E)$. If

a node is labeled by '*' its child is linked to the child of the corresponding node
in $TF(E)$. Such links are called *follow links*.

Notice that a follow link encodes the cartesian product of a Last set by a
First set, and that the transition relation δ is a union of such cartesian products.
Two products are either disjoint, or included in each other. Redundant products
are eliminated by a recursive procedure. Finally, the representation $ZPC(E)$ is
such that each transition is encoded in a unique follow link.

Example 1. Let take the expression $E = a(b+a)^* + a$. The linearized expression
is $E' = a_1(b_2 + a_3)^* + a_4$. So we can build the ZPC-representation as shown in
fig. 1.

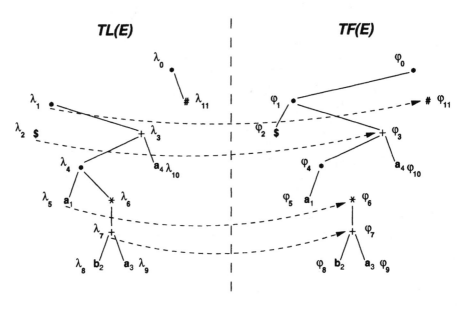

Fig. 1. ZPC(E) for $E' = a_1(b_2 + a_3)^* + a_4$.

It is convenient to process the expression $E' = \$(a_1(b_2 + a_3)^* + a_4)\#$, where $\$$
and $\#$ are two distinguished positions. The position $\$$ is associated to the initial
state of $\mathcal{M}_E$ and is involved in the follow link: $\{\$\} \times First(E)$. The position $\#$
is reached from positions which belong to $Last(E)$ by scanning the end of the
input word ; it appears only in the follow link $Last(E) \times \{\#\}$. Notice that $\$$ is
involved in this link too if $\$ \in Last(E)$, i.e. if $L(E)$ contains ε.

3   The Subset Construction

In this section we recall the subset construction.

Let $\mathcal{M} = (Q, \Sigma, \delta, I, F)$ be an arbitrary nondeterministic automaton. Let
$n = |Q|$. The subset construction [7,19] computes a deterministic automaton

$\mathcal{D} = (Q', \Sigma, \delta', \{0'\}, F')$ which recognizes the same language as $\mathcal{M}$. In order to deduce $\mathcal{D}$ from $\mathcal{M}$, we consider the application β which maps the states of $\mathcal{D}$ into the subsets of states of $\mathcal{M}$.

1. Let $\beta : Q' \longrightarrow 2^Q$ be a map such that:
   (a) $\beta(0') = I$
   (b) $\forall q' \in Q', \forall l \in \Sigma$, the state $q'_l = \delta'(q', l)$ is such that:

$$\beta(q'_l) = \bigcup_{q \in \beta(q')} \delta(q, l)$$

   (c) $\beta(q') = \emptyset \Rightarrow q' = s'$, where s' is a unique sink state of $\mathcal{D}$.
2. $q' \in F' \Leftrightarrow \beta(q') \cap F \neq \emptyset$

Notice that subset construction leads to two classical implementations:

1. The *exhaustive algorithm* first computes the transitions of all the $2^n - 1$ possible states and then trims the automaton.
2. The *reachable algorithm* only computes the transitions of the reachable states of $\mathcal{D}$. The difficulty is to decide whether the state $q'_l = \delta'(q', l)$ is a new state. This state identification test is generally based on comparisons between the set $\beta(q'_l)$ associated with q'_l and the sets $\beta(p')$ associated with states p' already in $\mathcal{D}$.

4   Determinization of Glushkov Automata

Glushkov automata have some nice properties [3] as do their ZPC-representation [15]. We take advantage of these properties, to improve the efficiency of subset construction applied to Glushkov automata.

Let us consider a regular expression E, and let $||E||$ be its alphabetic width and $\mathcal{M}_E = (Q, \Sigma, \delta, \{0\}, F)$ be its Glushkov automaton. We have $|Q| = n = ||E|| + 1$. Let $\mathcal{D}_E = (Q', \Sigma, \delta', \{0'\}, F')$ be the result of the determinization of $\mathcal{M}_E$ with $|Q'| = n'$. For each $l \in \Sigma$, we denote by n_l the number of occurrences of l in E and by Q_l the set of states q such that $\chi(q) = l$.

As far as Glushkov automata are concerned, we can reduce the complexity of subset construction as we now establish.

4.1   Bounding the Number of States in the Deterministic Automaton

Proposition 1. *The number n' of states of the deterministic automaton $\mathcal{D}_E$ yielded by the determinization of a Glushkov automaton satisfies*

$$n' \leq \left(\sum_{a \in \Sigma} 2^{n_a} \right) - |\Sigma| + 1.$$

Proof. This follows from the *homogeneity* of a Glushkov automaton; that is,

$$\forall q, p \in Q, \forall a, b \in \Sigma \ : \ \delta(q,a) = \delta(p,b) \Rightarrow a = b. \tag{1}$$

Let us consider a subset $P \subseteq Q$ ($P \neq \{0\}$) produced by the subset construction. There exists a subset S of Q and a letter a of Σ such that:

$$P = \bigcup_{s \in S} \delta(s,a).$$

Using (1) we have $P \subseteq \{ q \mid q \in Q \text{ and } \chi(q) = a \} = Q_a$ which shows that only subsets of $Q_{l \in \Sigma}$ have to be considered.

Remark 1. This bound is generally much smaller than the general bound $2^n - 1$. The gap increases with the size of the alphabet, which is generally large in linguistic applications.

Example 2. Consider the expression

$$E = (a_1 + (a_2 + b_3)^* a_4)(a_5 + b_6)^*.$$

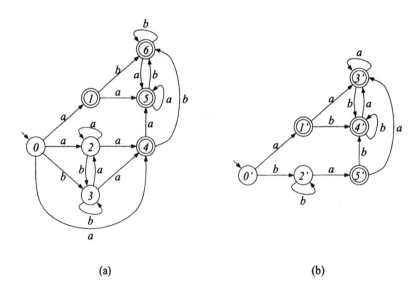

(a)   (b)

Fig. 2. a. The nondeterministic automaton $\mathcal{M}_E$. b. The deterministic automaton $\mathcal{D}_E$ of $\mathcal{M}_E$.

$\mathcal{M}_E$ and $\mathcal{D}_E$ are given by Figure 2. The number of states of $\mathcal{D}_E$ is at most 127 by the classical worst case bound and is at most 19 by Proposition 1.

Example 3. If we consider the expression

$$E = (a + b)^*(babab(a + b)^*bab + bba(a + b)^*bab)(a + b)^*$$

given by Antimirov [2], the classical bound is 8.3 million and our bound is 8.7 thousand.

A practical interest of our bound is to help in deciding whether an exhaustive implementation is practicable or not.

4.2   Computing the Transitions of the Deterministic Automaton

We now deal with the computation of the transitions of $\mathcal{D}_E$. Let q' be a state of $\mathcal{D}_E$ and l a letter of Σ. Let $q'_l = \delta'(q', l)$. We consider the set $\gamma(q') = \bigcup_{l \in \Sigma} \beta(q'_l)$.

As a Glushkov automaton is homogeneous, we have:

$$\forall q, r \in \beta(q'), \;\; \chi(q) = \chi(r),$$

for all states q' in Q'. So, for $l \in \Sigma$, the subsets $\beta(q'_l)$ associated to states q'_l are pairwise disjoint. Hence the set $\gamma(q')$ can be computed as a disjoint union.

We now show that the set $\gamma(q')$ can be computed in linear time, if the ZPC structure is used judiciously. We have

$$\gamma(q') = \biguplus_{l \in \Sigma} \bigcup_{s \in \beta(q')} \delta(s, l) = \biguplus_{l \in \Sigma} \delta(\beta(q'), l). \tag{2}$$

Let us give an efficient procedure which computes $Y = \biguplus_{l \in \Sigma} \delta(X, l)$, where $X \subseteq Q$. This procedure is detailed in the paper [14] which describes an efficient test for regular language membership. We briefly recall it here. It is based on three steps:

Step 1: Let us compute the set Λ of nodes of the forest $TL(E)$ which are at the same time ancestors of at least one $x \in X$, and heads of follow links. This is done by inspecting the path going from any position x of X to the root of the tree x belongs to.

Step 2: Let Φ be the set of tails of follow links associated to Λ.

Step 3: Given such a set Φ, Y is the set of positions of $TF(E)$ which are descendant of at least one element of Φ. Y is derived from the set Φ' of nodes of $TF(E)$ which are tails of follow links, such that:

$$Y = \biguplus_{\varphi \in \Phi'} First(\varphi)$$

Example 4.
  step 1 $X = \{1, 3, 4\}$ $\rightarrow \Lambda = \{\lambda_1, \lambda_5, \lambda_7\}$
  step 2 $\Lambda = \{\lambda_1, \lambda_5, \lambda_7\}$ $\rightarrow \Phi = \{\varphi_{11}, \varphi_6, \varphi_7\}$
  step 3 $\Phi = \{\varphi_{11}, \varphi_6, \varphi_7\}$ $\rightarrow \Phi' = \{\varphi_{11}, \varphi_6, \varphi_7\} \rightarrow Y = \{2, 3, \#\}$

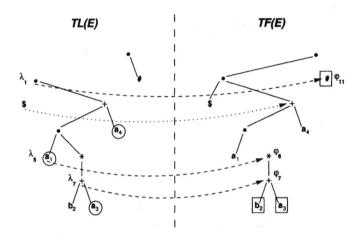

Fig. 3. ZPC(E) for $E' = a_1(b_2 + a_3)^* + a_4$.

Proposition 2 ([14]). *Let* $X \subseteq Q$ *and* $l \in \Sigma$. *Then, the set* $Y = \biguplus\limits_{l \in \Sigma} \delta(X, l)$ *can be computed in time* $O(n)$.

Proposition 3. *For all* $q' \in Q'$, $\gamma(q')$ *can be computed in time* $O(n)$.

Proof. From Formula (2) and Proposition 2.

4.3   Testing Set Equalities

The aim now is to reduce the number of integer comparisons involved by the test "Is q'_l a new state?". We are going first to show that we can use sets $\gamma\gamma(q')$ instead of sets $\gamma(q')$, with $|\gamma\gamma(q')| \leq |\gamma(q')|$, in order to improve the time complexity of the set equality test. Then we shall describe a linear time algorithm which computes the sets $\gamma\gamma(q')$. Let us explain how these reduced sets derive from $\gamma(q')$ sets.

Let us consider the set $\Phi'_{q'}$ which is such that $\gamma(q') = \biguplus\limits_{\varphi \in \Phi'_{q'}} First(\varphi)$.

The number of *First* sets which occur in this disjoint union is necessarily at most the number of positions in $\gamma(q')$. Therefore it seems conceivable to characterize the state q' by the set $\Phi'_{q'}$ and the state q'_l by the set $\{\varphi | \varphi \in \Phi'_{q'}$ and $First(\varphi) \cap Q_l \neq \emptyset\}$. This last set is smaller than $\beta(q'_l)$. However the set $\Phi'_{q'}$ is not a good candidate since for two distinct states q' and q'', one can have both $\gamma(q') = \gamma(q'')$ and $\Phi'_{q'} \neq \Phi'_{q''}$. This is shown by the following example.

Example 5. On the figure 3 of the example 4. After determinization, we get:

$$
\begin{array}{llll}
0' \rightarrow \{0\} & \gamma(0') = \{1, 4\} & \beta(0'_a) = \{1, 4\} & \beta(0'_b) = \emptyset \\
1' \rightarrow \{1, 4\} & \gamma(1') = \{2, 3, \#\} & \beta(1'_a) = \{3\} & \beta(1'_b) = \{2\} \\
2' \rightarrow \{3\} & \gamma(2') = \{2, 3, \#\} & & \\
3' \rightarrow \{2\} & \gamma(3') = \{2, 3, \#\} & &
\end{array}
$$

We have: $\gamma(1') = \gamma(2') = \{2, 3, \#\}$, and $\Phi'_{1'} = \{\varphi_6, \varphi_{11}\} \neq \Phi'_{2'} = \{\varphi_7, \varphi_{11}\}$.

Let us consider the following property:

$$\varphi \in TF(E),\ \mathcal{P}(\varphi) : \forall x \in Pos(\varphi), \exists \varphi' \in \Phi'_{q'} \mid x \in First(\varphi').$$

By convention, $\mathcal{P}(father(r))$ is false, if r is the root of a tree of $TF(E)$. We now consider sets $\gamma\gamma(q')$ such that:

$$\forall \varphi \in \gamma\gamma(q'),\ \mathcal{P}(\varphi) \wedge \neg\mathcal{P}(father(\varphi)).$$

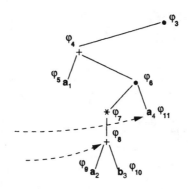

Fig. 4. Computation of $\gamma\gamma(q')$.

Example 6. Let us assume that: $\Phi'_{q'} = \{\varphi_8, \varphi_{11}\}$. We have: $\gamma\gamma(q') = \{\varphi_6\}$ since

1. $Pos(\varphi_6) = First(\varphi_8) \cup First(\varphi_{11})$
2. $\varphi_4 = father(\varphi_6)$ is such that a_1 belongs to $Pos(\varphi_4)$ while a_1 neither belongs to $Pos(\varphi_8)$ nor to $Pos(\varphi_{11})$.

Such a set $\gamma\gamma(q')$ is uniquely defined, since every position in $\biguplus\limits_{\varphi \in \Phi'_{q'}} First(\varphi)$ is represented by a unique ancestor, the closest one to the root. Hence the following proposition.

Proposition 4. *If $q' \neq q''$ and $\gamma(q') \neq \gamma(q'')$, then $\gamma\gamma(q') \neq \gamma\gamma(q'')$.*

Let us remark that since every state p' $(p' \neq 0')$ of Q' is generated by the action of a letter l on a state q' already generated, every state $(p' = q'_l)$ of Q' can be characterized by the set

$$\beta\beta(q'_l) = \{\varphi \mid \varphi \in \gamma\gamma(q') \text{ and } First(\varphi) \cap Q_l \neq \emptyset\}.$$

As $|\gamma\gamma(q'_l)| \leq |\gamma(q'_l)|$, we have $|\beta\beta(q'_l)| \leq |\beta(q'_l)|$, which speeds up the set equality tests.

```
procedure create-γγ(Φ', TF(E), γγ)
  /* Φ' and TF(E) are input parameters */
  /* γγ is an output parameter */
  /* d is a local boolean array */
  function dans(φ, in)
    /* φ is a node of T(E) used as input */
    /* in is a boolean array used as input/output */
    begin
      if (in[φ] = false) or (φ ∉ Pos(E)) then
        switch φ
        begin
          case '.','+' : in[φ] ← (dans(left(φ), in) ∧ dans(right(φ), in))
            case '*' : in[φ] ← dans(child(φ), in)
        end
      fi
      return in[φ]
    end
  procedure traversal(φ, d, γγ)
    /* φ is a node of T(E) used as input */
    /* γγ is an output parameter */
    begin
      if d[φ]
      then γγ ← γγ ∪ {φ}
      else if φ ∉ Pos(E)
          then
           begin
             traversal(left(φ), γγ)
             traversal(right(φ), γγ)
           end
      fi
    end
begin
  foreach φ ∈ TF(E) do d[φ] ← false
  foreach φ ∈ Φ' do
    if father(φ) = '*' then d[φ] ← true
  od
  dans(rac(T(E)), d)
  traversal(rac(T(E)), d, γγ)
end
```

Fig. 5. Algorithm for computation of sets $\gamma\gamma$.

We finally describe a linear time algorithm (see Figure 5) which computes the sets $\gamma\gamma(q')$.

The procedure *create-γγ* generates the set $\gamma\gamma(q')$. It calls the function *dans* and the procedure *traversal*. The function *dans* constructs an array *in* such that $in[\varphi] = \texttt{true}$ if and only if $\varphi \in \gamma\gamma(q')$. The procedure *traversal* collects the nodes of $\gamma\gamma(q')$ through a recursive traversal of $T(E)$.

5   Conclusion

The representation $\text{ZPC}(E)$ of the Glushkov automaton of a regular expression E, which can be computed in linear time and space, encodes a natural partition of the set of transitions. Making use of this partition speeds up the computation of the transitions and of the set equality tests in the subset construction when applied to Glushkov automata.

References

1. A. V. Aho, J. E. Hopcroft, and J. D. Ullman. *The design and analysis of computer algorithms.* Addison-Wesley, Reading, MA, 1974.
2. V. Antimirov. Partial derivatives of regular expressions and finite automaton constructions. *Theoret. Comput. Sci.*, 155:291–319, 1996.
3. P. Caron and D. Ziadi. Characterization of Glushkov automata. *Theoret. Comput. Sci.* to appear.
4. J.-M. Champarnaud and G. Hansel. Automate, a computing package for automata and finite semigroups. *J. Symbolic Comput.*, 12:197–220, 1991.
5. M.D. Davis, R. Sigal, and E.J. Weyuker. *Computability, Complexity, and Languages, Fundamentals of Theoretical Computer Science.* Academic Press, 1994.
6. V. M. Glushkov. The abstract theory of automata. *Russian Mathematical Surveys*, 16:1–53, 1961.
7. J. E. Hopcroft and J. D. Ullman. *Introduction to Automata Theory, Languages and Computation.* Addison-Wesley, Reading, MA, 1979.
8. J. H. Johnson. A program for computing finite automata. Unpublished Report, University of Waterloo, Canada, 1986.
9. J. H. Johnson and D. Wood. Instruction computation in subset construction. In D. Raymond, D. Wood, and S. Yu, editors, *Automata Implementation : First International Workshop on Implementing Automata, WIA'96*, number 1260 in Lecture Notes in Computer Science, pages 64–71, London, Ontario, 1997. Springer-Verlag, Berlin.
10. S. Kleene. Representation of events in nerve nets and finite automata. *Automata Studies*, Ann. Math. Studies 34:3–41, 1956. Princeton U. Press.
11. T. K. S. Leslie. Efficient approaches to subset construction. Technical Report CS-92-29, Department of Computer Science, University of Waterloo, Waterloo, Ontario, Canada, 1992.
12. T. K. S. Leslie, D. R. Raymond, and D. Wood. The expected performance of subset construction. non publié, 1996.
13. D. Perrin. Finite automata. In J. van Leeuwen, editor, *Handbook of Theoretical Computer Science, Formal Models and Semantics*, volume B, pages 1–57. Elsevier, Amsterdam, 1990.
14. J.-L. Ponty. An efficient null-free procedure for deciding regular language membership. In D. Wood and S. Yu, editors, *Automata Implementation : Second International Workshop on Implementing Automata, WIA'97*, number 1436 in Lecture Notes in Computer Science, pages 159–170, London, Ontario, 1998. Springer-Verlag, Berlin.

15. J.-L. Ponty, D. Ziadi, and J.-M. Champarnaud. A new quadratic algorithm to convert a regular expression into an automaton. In D. Raymond, D. Wood, and S. Yu, editors, *Automata Implementation : First International Workshop on Implementing Automata, WIA '96*, number 1260 in Lecture Notes in Computer Science, pages 109–119, London, Ontario, 1997. Springer-Verlag, Berlin.
16. M. O. Rabin and D. Scott. Finite automata and their decision problems. *IBM J. Res.*, 3(2):115–125, 1959.
17. D. Raymond and D. Wood. Grail, a C++ library for automata and expressions. *J. Symbolic Comput.*, 17:341–350, 1994.
18. D. Wood. *Theory of Computation.* Wiley, New York, 1987.
19. S. Yu. Regular languages. In G. Rozenberg and A. Salomaa, editors, *Handbook of Formal Languages*, volume I, Word, Language, Grammar, pages 41–110. Springer-Verlag, Berlin, 1997.

Implementing Reversed Alternating Finite Automaton (r-AFA) Operations*

Sandra Huerter, Kai Salomaa, Xiuming Wu, and Sheng Yu

Department of Computer Science
University of Western Ontario
London, Ontario, Canada N6A 5B7
{huerter,ksalomaa,wu6,syu}@csd.uwo.ca

Abstract. In [17], we introduced a bit-wise representation of r-AFA, which greatly improved the space efficiency in representing regular languages. We also described our algorithms and implementation methods for the union, intersection, and complementation of r-AFA. However, our direct algorithms for the star, concatenation, and reversal operations of r-AFA would cause an exponential expansion in the size of resulting r-AFA for even the average cases. In this paper, we will design new algorithms for the star, concatenation, and reversal operations of r-AFA based on the bit-wise representation introduced in [17]. Experiments show that the new algorithms can significantly reduce the state size of the resulting r-AFA. We also show how we have improved the DFA-to-AFA transformation algorithm which was described in [17]. The average run time of this transformation using the modified algorithm has improved significantly (by 97 percent).

1 Introduction

The study of finite automata was motivated largely by the study of control circuits and computer hardware in the fifties and early sixties. Implementation of finite automata was mainly a hardware issue then.

Since the mid and late sixties, finite automata have been widely used in lexical analysis, string matching, etc. They have been implemented in software rather than hardware. However, the sizes of the automata in those applications are small in general.

Recently, finite automata and their variants have been used in many new software applications. Examples are statecharts in object-oriented modeling and design [11,12,16,9], weighted automata in image compression [6], and synchronization expressions and languages in concurrent programming languages [10]. Many of those new applications require automata of a very large number of states. For example, concatenation is a required operation in most of those applications. Consider the concatenation of two deterministic finite automata (DFA)

---

* This research is supported by the Natural Sciences and Engineering Research Council of Canada grants OGP0041630.

with 10 states and 20 states, respectively. The resulting DFA may contain about 20 million states in the worst case [21].

It is clear that implementing finite automata in hardware is different from that in software. Hardware implementation is efficient, but suitable only for predefined automata. Adopting hardware implementation methods in software implementations is not immediate. It has to solve at least the following problems: (1) How do we represent and store a combinational network in a program efficiently in both space and time? (A table would be too big.) (2) A basic access unit in a program is a word. The access is parallel within a word and sequential among words. It would not be efficient if an implementation algorithm does not consider the word boundary and make use of it.

Implementing a small finite automaton is also different from implementing a very large finite automaton. A small finite automaton can be implemented by a word- based table or even a case or switch statement. However, these methods would not be suitable for implementing a DFA of 20 million states.

In [17], we introduced a bit-wise representation of r-AFA, which greatly improved the space efficiency in representing regular languages. We also described our algorithms and implementation methods for the union, intersection, and complementation of r-AFA.

It has been shown that a language L is accepted by an n-state DFA if and only if the reversal of L, i.e., L^R, is accepted by a $\log n$-state AFA. So, the use of r-AFA (reversed AFA) instead of DFA guarantees a logarithmic reduction in the number of states. However, the boolean expressions that are associated with each state can be of exponential size in the number of states. In our previous paper [17], we introduced a bit-wise representation for r-AFA and described the transformations between DFA and r-AFA, and also the algorithms for the union, intersection, and complementation for r-AFA. The model of r-AFA is naturally suited for bit-wise representations. NFA and DFA could also be represented in certain bit-wise forms which would save space. However, their operations would be awkward and extremely inefficient. Our experiments have shown that the use r-AFA instead of DFA for implementing regular languages can significantly improve, on average, both the space efficiency and the time efficiency for the union, intersection, and complementation operations.

We know that the resulting DFAs of the reversal and star operations of an n- state DFA have 2^n and $2^{n-1} + 2^{n-2}$ states, respectively, in the worst case, and the result of a concatenation of an m-state DFA and an n-state DFA is an $m2^n - 2^{n-1}$-state DFA in the worst case [21]. Therefore, in the worst case, the resulting r-AFA of the corresponding operations of r-AFA have basically the same state complexities, respectively. Direct constructions for the reversal, star, and concatenation of r-AFA, as described for AFA in [8,7], would have an exponential expansion in the number of states for each of the above mentioned operations.

In this paper, we present our new algorithms for the reversal, star, and concatenation operations of r-AFA. These algorithms simplify the r-AFA during the operations. They do not necessarily produce a minimum r-AFA, but they reduce the number of states tremendously in the average case.

Our experiments show that the algorithms reduce not only the size of the state set but also the total size of a resulting r-AFA in the average case.

At the end, we also show how we have improved the DFA-to-AFA transformation algorithm described in [17]. The average run time of this transformation using the improved algorithm has been reduced significantly (by 97 percent), and for the same input DFA, the resulting AFA of the original and the improved algorithms are the same up to a permutation of states.

2   Preliminaries

The concept of alternating finite automata (AFA) was introduced in [3] and [2] at the same period of time under different names. A more detailed treatment of AFA operations can be found in [8]. In the paper [17], we modified the definition of AFA and introduced h- AFA and r-AFA. The notion of "reversed" AFA, r-AFA, is considered also in [21] but the definition there differs in a minor technical detail from the definition used here and in [17]. Below we briefly recall the definition of AFA. For a more detailed exposition and examples the reader is referred to [21]. Background on finite automata in general can be found in [19].

We denote by B the two-element Boolean algebra and B^Q stands for the set of all functions from the set Q to B.

An h-AFA A is a quintuple (Q, Σ, g, h, F), where Q is the finite set of states, Σ is the input alphabet,

$$g : Q \times \Sigma \times B^Q \to B$$

is the transition function,

$$h : B^Q \to B$$

is the accepting Boolean function, and $F \subseteq Q$ is the set of final states.

We use $g_q : \Sigma \times B^Q \to B$ to denote that g is restricted to state q, i.e., $g_q(a, u) = g(q, a, u)$, for $a \in \Sigma, u \in B^Q, q \in Q$.

We also use g_Q to denote the function from $\Sigma \times B^Q$ to B^Q that is obtained by combining the functions $g_q, q \in Q$, i.e.,

$$g_Q \equiv (g_q)_{q \in Q}.$$

We will write g instead of g_Q whenever there is no confusion.

Let $u \in B^Q$. We use $u(q)$ or $u_q, q \in Q$, to denote the component of the vector u indexed by q. By $\mathbf{0}$ we denote the constant zero-vector in B^Q (when Q is understood from the context).

We extend the definition of g of an h-AFA to a function: $Q \times \Sigma^* \times B^Q \to B$ as follows:

$$g(q, \lambda, u) = u_q$$
$$g(q, aw, u) = g(q, a, g(w, u))$$

for all $q \in Q, a \in \Sigma, \omega \in \Sigma^*, u \in B^Q$.

Similarly, the function g_Q, or simply g, is extended to a function $\Sigma^* \times B^Q \rightarrow B^Q$.

Given an h-AFA $A = (Q, \Sigma, g, h, F)$, for $w \in \Sigma^*$, w is accepted by A if and only if $h(g(w, f)) = 1$, where f is the characteristic vector of F, i.e., $f_q = 1$ iff $q \in F$.

An r-AFA A is an h-AFA such that for each $w \in \Sigma^*$, w is accepted by A if and only if $h(g(w^R, f)) = 1$, where f is the characteristic vector of F.

The transition functions of h-AFA and r-AFA are denoted by Boolean functions. Every Boolean function can be written as a disjunction of conjunctions of Boolean variables. We call a conjunction of Boolean variables or a constant Boolean function a term. A term $x_1 \wedge \ldots \wedge x_k$ is denoted simply as $x_1 \cdots x_k$. The negation of a variable x is denoted as $\bar{x}$.

The following result, that was proved in [17] states that any Boolean term can be represented by two bit-wise vectors.

Theorem 1 *For any Boolean function f of n variables that can be expressed as a single term, there exist two n-bit vectors α and β such that*

$$f(u) = 1 \iff (\alpha \& u) \uparrow \beta = \mathbf{0}, \quad \text{for all } u \in B^n,$$

where $\&$ is the bit-wise AND operator, $\uparrow$ the bit-wise EXCLUSIVE-OR operator, and $\mathbf{0}$ is the zero vector $(0, \ldots, 0)$ in B^n.

Each n-bit vector $v = (v_1, \ldots, v_n)$, $n \le 32$ (32-bit is the normal size of a word), can be represented as an integer

$$I_v = \sum_i^n v_i 2^{i-1}.$$

We can also transform an integer I_v back to a 32-bit vector v in the following way:
$$v_i = (I_v \& 2^{i-1})/2^{i-1}, \quad 1 \le i \le n.$$

So, a Boolean function, which is in disjunctive normal form, can be represented as a list of terms, while each term is represented by two integers. For an r-AFA $A = (Q, \Sigma, g, h, F)$, where $Q = \{q_1, \ldots, q_n\}$ and $\Sigma = \{a_1, \ldots, a_m\}$, we can represent g as a table of functions of size $n \times m$ with (i, j) entry corresponding to the function $g_{q_i}(a_j) : B^Q \rightarrow B$ defined by $[g_{q_i}(a_j)](u) = g_{q_i}(a_j, u)$, for $q_i \in Q$, $a_j \in \Sigma$, and $u \in B^Q$. The accepting function h can be represented as a list of integer pairs. Finally, F can be represented as a bit-vector (an integer), i.e., as its characteristic vector.

3   Algorithms for Operations of r-AFA

In this section, we present the algorithms for constructing the star and reversal of a given r-AFA and the concatenation of two given r-AFA. Note that after each construction, a simplification algorithm is applied to each Boolean function in order to reduce the size of the Boolean expression. An algorithm which we have implemented can be found in [20]. We omit the proofs of the correctness of the algorithms due to the limit on the size of the paper.

3.1   The Star Operation

First, we consider the star operation. The algorithm eliminates all the useless states during the construction. Our experiments suggest that the r-AFA resulting from this method are significantly smaller than the naive algorithm described in [8].

Let $A = (Q, \Sigma, g, h, F)$ be a given r-AFA with $Q = \{q_1, \ldots, q_n\}$. We construct a r-AFA $A' = (Q', \Sigma, g', h', F')$ such that $L(A') = L(A)^*$ in the following. Let f be the characteristic vector of F.

Now we describe the algorithm, which deletes all the useless states during the construction.

If $n = 0$ and $h = 1$, then the r-AFA A accepts Σ^*. So, $L(A') = (L(A))^*$ is the same as $L(A)$ and we can just let $A' = A$. If $n = 0$ and $h = 0$, then A accepts the empty language and A' accepts the language that contains only the empty word λ. Then A' is the one-state r-AFA where $Q' = \{0\}$, $F' = \emptyset$, $h(x_0) = \bar{x}_0$, and $g(a, x_0) = 1$ for all $a \in \Sigma$.

Now we assume that n is a positive integer.

$F' = \emptyset$.
Let I be an array of integers of size 2^n.
Initialize $I[k] = 0$, for $k = 0, \ldots, 2^n - 1$.
Use the procedure Markarray(I, A) described at the end of this subsection to mark I, so that
$I[k] = 1 \iff \exists x \in \Sigma^*$, s.t. $g(x, f) = k$, for $k \neq f$;
$I[f] = 1 \iff \exists x \in \Sigma^*$ s.t. $h(g(x, f)) = 1$.
If $I[f] = 0$, that means that A accepts nothing. We can construct A' the same way as in the case $n = 0$. Otherwise go to the next steps.
Let $I[k_0], \ldots, I[k_p]$ be all the entries of I that have a value 1. Let P be an array of integers of size p, such that $P[i] = k_i$.
Set $Q' = \{0, \ldots, p\}$.
Similarly as above we assume that the set $\{0, 1, \ldots, 2^n - 1\}$ is identified with B^Q. This means that, for all $i \in Q'$, the entry $P[i]$ will represent an element of B^Q.
Define the head function h' as follows:

$$h'(u) = 1 \iff u = \mathbf{0} \quad \text{or} \quad \exists t \in Q' \text{ such that } u_t = 1, \ h(P[t]) = 1.$$

Thus,

$$h'((x_0, \ldots, x_p)) = \bar{x}_0 \ldots \bar{x}_p \vee \bigvee_{h(P[i])=1} x_i.$$

For any $a \in \Sigma$ and k such that $P[k] \neq f$, define g'_k as:

$$g'(k, a, u) = 1 \iff \exists s \in Q', \quad \text{such that} \quad u_s = 1, \quad g(a, P[s]) = P[k]$$

if $u \neq \mathbf{0}$; and

$$g'(k, a, \mathbf{0}) = 1 \iff g(a, f) = P[k].$$

That is,

$$g'(k, a, (x_0, \ldots, x_p)) = \begin{cases} \bigvee_{g(a, P[i]) = P[k]} x_i \vee \bar{x}_0 \ldots \bar{x}_p & \text{if } g(a, f) = P[k] \\ \bigvee_{g(a, P[i]) = P[k]} x_i & \text{otherwise.} \end{cases}$$

Assume t is an index such that $P[t] = f$. Define g'_t as:

$$g'(t, a, u) = 1 \iff \exists s \in Q' \text{ such that } u_s = 1, \quad g(a, P[s]) = f$$
$$\text{or } \exists r \in Q', \text{ such that } u_r = 1, \quad h(g(a, P[r])) = 1$$

if $u \neq \mathbf{0}$; and

$$g'(t, a, \mathbf{0}) = 1 \iff g(a, f) = f \quad \text{or} \quad h(g(a, f)) = 1.$$

Thus, $g'(t, a, (x_1, \ldots, x_p)) =$

$$\begin{cases} \bigvee_{g(a, P[i]) = f \text{ or } h(g(a, P[i])) = 1} x_i \vee \bar{x}_0 \ldots \bar{x}_p & \text{if } h(g(a, f)) = 1 \text{ or } g(a, f) = f \\ \bigvee_{g(a, P[i]) = f \text{ or } h(g(a, P[i])) = 1} x_i & \text{otherwise.} \end{cases}$$

The following procedure can be used to mark the array used in the above algorithm.

Procedure Markarray(I, A)

Qu is a queue of integers. Initially, Qu has only one entry f.

```
if (h(f) == 1)
    I[f] = 1;
while (!Empty(Qu)
    {
    int tmp = pop (Qu);
    int vector;
    for (a ∈ Σ)
        {
        vector = g(a, tmp);
        if(I[vector] == 0)
            {
            I[vector] = 1;
            push(vector, Qu);
            }
        if (I[f] == 0 && h(vector) == 1)
            I[f] = 1;
        }
    }
```

3.2   Concatenation of r-AFA

Let $A^{(i)} = (Q^{(i)}, \Sigma^{(i)}, g^{(i)}, h^{(i)}, F^{(i)})$, where $i = 1, 2$, be two r-AFAs. We construct a r-AFA $A = (Q, \Sigma, g, h, F)$ which accepts the concatenation of $L(A^{(1)})$ and $L(A^{(2)})$.

Let the numbers of states of $A^{(1)}$ and $A^{(2)}$ are n and m, and f_i be the characteristic vectors of $F^{(i)}$ for $i = 1, \ 2$ respectively. We construct A in three cases as follows:

First, let us assume $mn \neq 0$ and $\Sigma^{(1)} = \Sigma^{(2)}$.

$Q = \{q_0, \ldots, q_{n-1}, q_n, \ldots, q_{n+2^m-1}\}$, where $Q^{(1)} = \{q_0, \ldots, q_{n-1}\}$ and q_k are new states for $k \geq n$.

$$F = \begin{cases} F^{(1)} & \text{if } h^{(1)}(f_1) = 0 \\ F^{(1)} \cup \{q_{n+f_2}\} & \text{otherwise.} \end{cases}$$

We identify the numbers $n, \ldots, n + 2^m - 1$ with elements of $B^{Q^{(2)}}$, thus in the above definition of F the notation $n + f_2$ stands for the number belonging to $\{n, \ldots, n + 2^m - 1\}$ that denotes f_2.

Define $h(u) = 1 \iff \exists x$, such that $0 \leq x \leq 2^m - 1$, $u_{x+n} = 1$, and $h^{(2)}(x) = 1$, for $u \in B^Q$; that is,

$$h((x_0, \ldots, x_{n+2^m-1})) = \bigvee_{h^{(2)}(i)=1} x_{n+i}.$$

We define $g(a, u)|_{Q^{(1)}} = g^{(1)}(a, u|_{Q^{(1)}})$, for $u \in B^Q$ and $a \in \Sigma$. That is,

$$g(q_i, a, (x_0, \ldots, x_{n+2^m-1})) = g^{(1)}(q_i, a, (x_0, \ldots, x_{n-1})) \ (\forall i < n).$$

Also define $g(a, u)_{q_x} = 1 \iff \exists \ y$, such that $0 \leq y \leq 2^m - 1$ and $u_{y+n} = 1$, $g^{(2)}(a, y) = x - n$, for $x \geq n$, $x \neq n + f_2$, $u \in B^Q$ and $a \in \Sigma$; that is,

$$g(q_i, a, (x_0, \ldots, x_{n+2^m-1}) = \bigvee_{g^{(2)}(a, k-n)=i-n} x_k.$$

Define $g(a, u)_{q_{n+f_2}} = 1 \iff h^{(1)}((g(a, u)|_{Q^{(1)}})) = 1$ or $\exists \ y$, such that $y \geq n$, $u_y = 1$, and $g^{(2)}(a, y - n) = f_2$, for all $u \in B^Q$ and $a \in \Sigma$. Thus,

$$g(q_{n+f_2}, a, (x_0, \ldots, x_{n+2^m-1}) = \bigvee_{g^{(2)}(a, k-n)=f_2-n} x_k \bigvee T(a, x_0, \ldots, x_{n-1}).$$

where the Boolean function T is the resulting function obtained by substituting $g^{(1)}(q_i, a, (x_0, \ldots, x_{n-1}))$ for x_i for all $i < n$.

Secondly, if $\Sigma^{(1)} \neq \Sigma^{(2)}$, we can construct two r-AFAs $A_1 = (Q_1, \Sigma, g_1, h_1, F_1)$ and $A_2 = (Q_2, \Sigma, g_2, h_2, F_2)$, where $\Sigma = \Sigma^{(1)} \cup \Sigma^{(2)}$, such that $L(A_i) = L(A^{(i)})$ for $i = 1, 2$. For example, let assume that $\Sigma \neq \Sigma^{(1)}$. The construction of A_1 is as follows:

Construct $Q_1 = Q^{(1)} \cup \{newq\}$, where $newq$ is a new state.
Set $F_1 = F^{(1)}$.
Define h_1 as: $h_1(u) = h^{(1)}(u|_{Q^{(1)}}) \wedge \bar{u}_{newq}$ for all $u \in B^{Q_1}$.
For any $q \in Q^{(1)}$, $a \in \Sigma$ and $u \in B^{Q_1}$, define:

$$g_1(q, a, u) = \begin{cases} g^{(1)}(q, a, u|_{Q^{(1)}}) & \text{if } a \in \Sigma^{(1)} \\ 0 & \text{otherwise.} \end{cases}$$

For any $a \in \Sigma$ and $u \in B^{Q_1}$, we define:

$$g_1(newq, a, u) = \begin{cases} u_{newq} & \text{if } a \in \Sigma^{(1)} \\ 1 & \text{otherwise.} \end{cases}$$

The last case is $mn = 0$. We can assume $\Sigma^{(1)} = \Sigma^{(2)}$ in this case (otherwise we can use the above construction to convert to this case).

Assume that $n = 0$ and $m \neq 0$. We omit the other cases here because they use a similar construction.

If $h^{(1)} = 0$, then $L(A^{(1)}) = \emptyset$. So, $L(A^{(1)})L(A^{(2)}) = \emptyset$. Therefore, we can let $A = A^{(1)}$
If $h^{(1)} = 1$, construct $A_1 = (Q_1, \Sigma, g_1, h_1, F_1)$, such that $|Q_1| = 1$ and $L(A_1) = L(A^{(1)})$ as follows:
    $Q_1 = \{q\}$, $h_1 = 1$ and $F_1 = \emptyset$, where q is a new state;
    $g_1(q, a, u) = 0$ for all $a \in \Sigma$, $u \in B^{Q_1}$.
Use the algorithm for the case $mn \neq 0$ and $\Sigma^{(1)} = \Sigma^{(2)}$ to construct the r-AFA accepting the concatenation of $L(A_1)$ and $L(A^{(2)})$.

Note that in the construction for the first case, we don't really need all the 2^m states that resulted from $A^{(2)}$. In fact, most of the states are of no use in general. We can use a similar method as was used in the construction for the star operation to reduce the state complexity. Briefly stated, we mark all the q_i for $i \geq n$ such that $i - n$ can be reached from f_2 under the $A^{(2)}$ transitions. Then we use the marked states to do the construction. We omit the construction here because the reader can easily fill in the details.

3.3   Reversal of r-AFA

The "reversal" of a r-AFA $A = (Q, \Sigma, g, h, F)$ is a r-AFA $A' = (Q', \Sigma, g', h', F')$ which accepts the reversal of the language of A, that is, $L(A') = [L(A)]^R$. Next we give the construction for the reversal operation, which simplifies the reversal r-AFA during the construction by removing all the useless and unreachable states. We assume $n > 0$.

Let I be an array of integers of size 2^n, Qu be a queue of integers, $T[2^n]$ be an array of integer pointers, $s = |\Sigma|$ and f be the characteristic vector of F.

Initially, let

Qu be empty;

$I[i] = 0$, for all $0 \le i < 2^n$;

$T[i] = 0$ for all $0 \le i < 2^n$;

int $index = 0$;

int $temp, vector$;

$Qu.\text{push}(f)$;

Find all vectors $v \in B^Q$ that can be reached by f, i.e., $\exists x \in \Sigma^*$ such that $v = g(x, f)$. The details are as follows:

```
while (!Qu.empty())
    {
    temp = Qu.pop();
    T[index] =new int[s + 1];
    T[index][0] = temp;
    I[temp] = index;
    for (a ∈ Σ)
        {
        vector = g(a, temp);
        T[index][a] = vector;
        if ((vector! = f) && (I[vector] == 0))
            Qu.push(vector);
        }
    index++;
    }
for (int i = 0; i < index; i + +)
    for (a ∈ Σ)
        T[i][a] = I[T[i][a]];
```

Construct an inverse table for T.

Let $R[index][s]$ be a two-dimensional array of sets of integers. The function $\text{Add}(R[i][j], k)$ add the integer k into the set $R[i][j]$. Initially, $R[i][j] = \emptyset$ for all $0 \le i < index$, $0 \le j < s$. Qu is also initialized to be empty.

```
for (int i = 0; i < index; i + +)
    {
    for (a ∈ Σ)
        Add(R[T[i][a]][a], i);
    if (h(T[i][0]) == 1)
        {
        Qu.push(i);
        Add(F'', i);       F'' is a set of integers
        }
    }
```

Mark the new useful states.

Initialize $I[k]$ to 0, for $k = 0, \ldots, index - 1$

if $(h(f) == 1)$ $I[0] = 1$;

while $(!Qu.\text{empty}())$

```
{
temp = Qu.pop();
if (!empty(R, temp))
    {
    I[temp] = 1;
    for (a ∈ Σ)
        for (q ∈ R[temp][a])
            if (I[q] == 0)
                Qu.push(q);
    }
}
```
Rename the states.
```
    if (I[0] == 0)
        A' accepts the empty language;
    else
        {
        int k = 0;
        for (int i = 0; i < index; i + +)
            if (I[i]! = 0)
                {
                I[i] = k;
                k + +;
                }
        }
```
Construct A'.
```
    Construct F'.
        Let F' = ∅;
        for (i in F'')
            if (I[i]! = 0)      Add(F', I[i])
    Construct the transition function g'.
        for (int i = 0; i < index; i + +)
            if (I[i]! = 0)
                for (a ∈ Σ)
```
$$g'(I[i], a) = \mathbf{0} \bigvee_{q \in R[i][a]} x_{I[q]};$$
```
    Construct the head function h'.
        h'(u) = u_0;
```

4 Improved Implementation of the DFA to r-AFA Transformation

In [17] it is shown that the bitwise representation of r-AFA significantly reduces the space needed to implement regular languages. However, manipulating bit-vectors is time-consuming if they are handled as arrays of 1's and 0's rather than integers. This is exemplified by the improvements made recently to the implementation of the DFA to r-AFA transformation of [20]. The majority of these improvements involve increasing the efficiency of bit-vector handling, and are described below. Overall the run-time of this transformation has been decreased by 97 percent. Thus, the computation of r-AFA for large input DFA is now entirely feasible with respect to time.

4.1   Handling Bit-Vectors: Weight

The computation of an r-AFA for a given 2^n-state DFA involves the set of bit-vectors $B^n = \{0, 1, 2, .., (2^n) - 1\}$. Several parts of the DFA to r-AFA transformation involve information related to the number of 1's contained in these bit-vectors (referred to as their "weight" in [20]):

Step 2a): given an interval of integers, find the integer of lowest weight;

Step 3b): sort arrays P_f and P_n of bit-vectors ascending on weight;

Step 7 : simplifying Boolean functions; two terms t1 and t2 differ only in the negation of of one variable iff $t1 \uparrow t2$ ($\uparrow$: Exclusive OR) has weight one.

These three parts of the transformation were in fact the most time-consuming, because computing the weight of elements of B^n was done directly:

```
for(i=0; i<32; i++)
if( bit_vector & pow(2,i) == pow(2,i) )
weight++;
```

Since the transformation requires such weight-related information, it has proven extremely useful to build an array W which contains the elements of B^n in order of increasing weight. The procedure Build_Weight_Array(int n) described below accomplishes this (in time O(n)). Traversing W is then the same as traversing B^n in order of increasing weight. Thus, steps 3a) and 3b) of the transformation can be implemented as one loop:

```
for each bit-vector u in W
if( h(u)==1 )
add u to array Pf;
else
add u to array Pn;
```

After filling P_f and P_n this way, they are already sorted ascending on weight (making implementation of step 3b) unnecessary). This improvement alone reduced the runtime of the filter by 70 percent.

To compare the weights of some subset S of B^n, an array W', the inverse of W, is useful, where $W'[W[i]] = i$ for all i in B^n. Then for any k in B^n W'[k] is the array index of k in W. So if k is the least element in set W'[S], then W[k] is the element of S with the smallest weight. Using this method to compute step 2a) also greatly reduced the overall runtime of the transformation.

Finally, one of the tests most often applied in the Boolean function simplification procedure being used is whether two terms differ only in the negation of one variable. And since we are representing terms using bit-vectors, two terms t1 and t2 differ only in the negation of one variable iff $t1 \uparrow t2$ has weight one.

Previously, all 32 positions of the bit-vector $t1 \uparrow t2$ were scanned for 1's. But if array W' is available, then

(the weight of bit-vector v is 1) iff (1 <= W'[v] <= n)

making only two integer comparisons necessary for the test. This improvement decreased the runtime of the simplification routine by 95 percent.

What follows is the pseudocode describing how array W is constructed. To simplify the description, W is described in terms of blocks, where block i contains the integers of B^n with weight i. Numbers in block $i + 1$ are generated by either incrementing (giving an "incremented_int") certain numbers in block i, or shifting (giving a "shifted_int") certain numbers in block $i + 1$ as follows:

```
Build_Weight_Array(n)
{
W[0] = 0;   //0 is considered a shifted_int

for(i = 1 .. n) {
for each shifted_int s in block i-1,
(next incremented_int of block i of W) := s + 1;

for each incremented_int j in block i
while( k=leftshift(j) <= (2^n)-1 )
(next shifted_int of block i of W) := k;
}
}
```

For example, let n=4. Then,

```
%
    block 0 : W[0] = ~   = 0 = 0000%
    block 1 : W[1] = 0+1     = 1 = 0001%
    W[2] = <<1   = 2 = 0010%
    W[3] = <<(<<1)    = 4 = 0100%
    W[4] = <<(<<(<<1)) = 8 = 1000%
    block 2 : W[5] = 2+1     = 3 = 0011%
    W[6] = 4+1   = 5 = 0101%
    W[7] = 8+1   = 9 = 1001%
    W[8] = <<3   = 6 = 0110%
    W[9] = <<(<<3)    = 12 = 1100%
    W[10]= <<5   = 10 = 1010%
    block 3 : W[11]= 6+1     = 7 = 0111%
    W[12]= 12+1 = 13 = 1101%
    W[13]= 10+1 = 11 = 1011%
    W[14]= <<7   = 14 = 1110%
    block 4 : W[15]= 14+1    = 15 = 1111%
```

References

1. J. Berstel and M. Morcrette, "Compact representation of patterns by finite automata", *Pixim 89: L'Image Numérique à Paris*, André Gagalowicz, ed., Hermes, Paris, 1989, pp.387-395.

2. J.A. Brzozowski and E. Leiss, "On Equations for Regular Languages, Finite Automata, and Sequential Networks", *Theoretical Computer Science* 10 (1980) 19-35.

3. A.K. Chandra, D.C. Kozen, L.J. Stockmeyer, "Alternation", *Journal of the ACM* 28 (1981) 114-133.

4. H.K. Cheung, *An Efficient Implementation Method for Alternating Finite Automata*, MSc Project Paper, Dept. of Computer Science, Univ. of Western Ontario, Sept. 1996.

5. Olivier Coudert, "Two-Level Logic Minimization: An Overview", Integration, *The VLSI Journal* 17 (1994) 97-140.

6. K. Culik II and J. Kari, "Image Compression Using Weighted Finite Automata", *Computer and Graphics*, vol. 17, 3, (1993) 305-313.

7. A.Fellah, *Alternating Finite Automata and Related Problems*. PhD dissertation, Kent State Univ. 1991.

8. A.Fellah, H.Jürgensen, and S.Yu, "Constructions for Alternating Finite Automata", *Intern J. Comp. Math.* 35 (1990) 117-132.

9. M.Fowler and K.Scott, *UML Distilled*, Addison-Wesley, 1997.

10. L. Guo, K. Salomaa, and S. Yu, "Synchronization Expressions and Languages", *Proceedings of the Sixth IEEE Symposium on Parallel and Distributed Processing* (1994) 257-264

11. D. Harel, "Executable Object Modeling with Statecharts", July 1997, *IEEE Computer*, pps. 31-42.

12. D. Harel, "Statecharts: a visual formalism for complex systems", *Science of Computer Programming* 8 (1987) 231-274.

13. H.B. Hunt, D.J. Rosenkrantz, and T.G. Szymanski, "On the Equivalence, Containment, and Covering Problems for the Regular and Context-Free Languages", *Journal of Computer and System Sciences* 12 (1976) 222-268.

14. T. Jiang and B. Ravikumar, "Minimal NFA Problems are Hard", *SIAM Journal on Computing* 22 (1993) 1117-1141.

15. D. Raymond and D. Wood, *Release Notes for Grail Version 2.5*, Dept. of Computer Science, Univ. of Western Ontario, 1996.

16. J. Rumbaugh, M. Blaha, W. Premerlani, F. Eddy, W. Lorensen, *Object-Oriented Modeling and Design*, Prentice-Hall, 1991.

17. K. Salomaa, X. Wu and S. Yu, "Efficient Implementation of Regular Languages Using R-AFA", *Proceedings of the Second International Workshop on Implementing Automata. Lecture Notes in Computer Science* 1436, Springer, pps.176-184.

18. L. Stockmeyer and A. Meyer, "Word problems requiring exponential time (preliminary report)", *Proceedings of the 5th ACM Symposium on Theory of Computing*, (1973) 1-9.

19. D. Wood, *Theory of Computation*, John Wiley & Sons, New York, 1987.

20. X. Wu, "Implementation of Regular Languages by Using R-AFA", *Master's Project Report*, The Department of Computer Science, Univ. of Western Ontario, 1997.

21. S. Yu, "Regular Languages", Chapter 2, *Handbook of Formal Languages*, Vol. 1, Springer 1997.

Operations on DASG*

Zdeněk Troníček

Department of Computer Science and Engineering
Czech Technical University
Karlovo nàm. 13
121 35 Prague 2, Czech Republic
tronicek@fel.cvut.cz

Abstract. Directed Acyclic Subsequence Graph (DASG) is an automaton that accepts all subsequences of a given string T. DASG allows us to decide whether a string S is a subsequence of T in $\mathcal{O}(|S|)$ time where $|S|$ is the length of S. We show that if we slightly modify the string T, it is possible to get the DASG for the modified string from the original DASG. For this purpose we define these operations on DASG: adding a state on the left, deleting a state on left, adding a state on the right, deleting a state on the right, adding an inner state, deleting an inner state and replacing a transition label. For each of these operations we describe the modification of DASG and the proof of correctness.

1 Introduction

A subsequence of a string is any string obtained by deleting zero or more symbols from the given string. Subsequences play an important role in data processing and genetic applications ([5]). For example, the longest common subsequence ([3]) and sequence alignment ([4]) are the best known problems. An important question to answer is the membership problem. That is, we are to determine whether a given string S is a subsequence of another string T. If we allow the string T to be preprocessed, we can answer the question in optimal time (that is, in time $\mathcal{O}(|S|)$). During the preprocessing of the string T we build an automaton that accepts all subsequences of this string. Such an automaton is called Directed Acyclic Subsequence Graph (DASG) and was introduced in [1]. A left-to-right algorithm for building DASG is described in [6]. DASG is analogous to Directed Acyclic Word Graph (DAWG)[2] using subsequences instead of substrings.

Let Σ be an alphabet and $T = t_1 t_2 \ldots t_n$ a string over this alphabet. The DASG for the string T is a finite automaton $A = (Q, \Sigma, \delta, q_0, F)$, where $Q = \{q_0, q_1, \ldots, q_n\}$ is a set of states, Σ is the input alphabet, $\delta : Q \times \Sigma \to Q$ is a transition function, q_0 is the initial state, and $F = Q$ is a set of final states (all the states are final). Clearly, the automaton has minimal number of states as it must accept the string T. The automaton can be partial, that is, each state need not have a transition defined for each symbol. The transition function δ is defined as follows (for $a \in \Sigma$ and $0 \le i \le n$):

---

* This research has been supported by GAČR grant No. 201/98/1155

J. M. Champarnaud, D. Maurel, D. Ziadi (Eds.): WIA'98, LNCS 1660, pp. 82–91, 1999.

$\delta(q_i, a) = q_j$, if there exists $k > i$ such that $a = t_k$ (j is minimal such k),
$\delta(q_i, a) = \emptyset$, otherwise.

For convenience, we define $\delta^* : Q \times \Sigma^* \to Q$ recursively by:
$\delta^*(q, a) = \delta(q, a)$
$\delta^*(q, xa) = \delta(\delta^*(q, x), a)$
for $q \in Q, a \in \Sigma$ and $x \in \Sigma^*$.

Let us suppose that we have built the DASG for the string $T = t_1 t_2 \ldots t_n$ and that the string T' can be obtained ¿from the string T using one of the following operations: adding a symbol, deleting a symbol, and replacing a symbol. We show how to modify the DASG for the string T to obtain the DASG for the string T'. For this purpose we describe these operations on the DASG:

- adding a state on the left,
- deleting a state on the left,
- adding a state on the right,
- deleting a state on the right,
- adding an inner state,
- deleting an inner state,
- replacing a transition label.

Obviously, adding a state on the left and on the right are special cases of adding an inner state. Similarly, deleting a state on the left and on the right are special cases of deleting an inner state. As the algorithms for these special cases are much simpler than for general operations, we present them separately. Each operation preserves the minimality of the number of states which results from the fact that the modified automaton have to accept the modified string T'.

For the time complexity, we consider representation of the DASG with explicit expression of transitions. That is, if we delete a state of the DASG, we have to delete all its transitions and it takes at least time linear in the number of transitions of this state.

2   Adding a State on the Left

The operation is required if we add a symbol x before the first symbol of the string T. Then, $T' = xT = xt_1 t_2 \ldots t_n$. The DASG for the string T' is $A' = (Q', \Sigma, \delta', q, Q')$, where $Q' = Q \cup \{q\}$ and δ' is defined as follows:
$\delta'(p, a) = \delta(p, a)$ for all $p \in Q$ and all $a \in \Sigma$,
$\delta'(q, x) = q_0$,
$\delta'(q, a) = \delta(q_0, a)$ for all $a \in \Sigma, a \neq x$.

Lemma 1 *A' accepts a string $S = s_1 s_2 \ldots s_m$ if and only if S is a subsequence of the string T'.*

Proof. We prove the Lemma in two steps.

1. A' accepts all subsequences of T': For any subsequence S of T' the following holds: $S = R$, or $S = xR$, where R denotes a subsequence of T. Let us consider both cases:

(a) $S = R$: If $s_1 \neq x$, then $\delta'(q, s_1) = \delta(q_0, s_1)$ and because S is accepted by A, it is accepted by A' as well.

If $s_1 = x$, then $\delta'(q, s_1) = q_0$. If $s_1 s_2 \ldots s_m$ is a subsequence of T, then $s_2 \ldots s_m$ is also a subsequence of T. Therefore, $s_2 \ldots s_m$ is accepted by A and S is accepted by A'.

(b) $S = xR$: $\delta'(q, x) = q_0$. As S is accepted by A, xS is accepted by A'.

2. If S is accepted by A', then S is a subsequence of T': Again, there are two possibilities:

(a) $s_1 = x$: $\delta'(q, s_1) = q_0$ and $s_2 \ldots s_m$ is accepted by A, that is, $s_2 \ldots s_m$ is a subsequence of T. Consequently, S is a subsequence of T'.

(b) $s_1 \neq x$: $\delta'(q, s_1) = \delta(q_0, s_1)$ and S is accepted by A, that is, S is a subsequence of T. Hence, S is a subsequence of T' as well.

□

Algorithm:
Input: DASG for the string $T = t_1 t_2 \ldots t_n$
Output: DASG for the string $T' = x t_1 t_2 \ldots t_n$

1: $\delta(q, x) \leftarrow q_0$
2: **for** each $a \in \Sigma, a \neq x$ **do**
3: $\quad \delta(q, a) \leftarrow \delta(q_0, a)$
4: **end for**
5: $Q \leftarrow Q \cup \{q\}$

Time complexity: the algorithm requires $\mathcal{O}(|\Sigma|)$ time.

3 Deleting a State on the Left

The operation is required if we delete the first symbol of the string T. Then, $T' = t_2 t_3 \ldots t_n$. The DASG for the string T' is $A' = (Q', \Sigma, \delta', q_1, Q')$, where $Q' = Q \setminus \{q_0\}$ and δ' is defined as follows:
$\delta'(p, a) = \delta(p, a)$ for all $p \in Q'$ and all $a \in \Sigma$.

Lemma 2 A' accepts a string $S = s_1 s_2 \ldots s_m$ if and only if S is a subsequence of the string T'.

Proof. We prove the Lemma in two steps.

1. A' accepts all subsequences of T': If S is a subsequence of T', then $t_1 S = t_1 s_1 s_2 \ldots s_m$ is a subsequence of T. Consequently, $t_1 S$ is accepted by A. As $\delta(q_0, t_1) = q_1$, S is accepted by A'.

2. If S is accepted by A', then S is a subsequence of T': If A' accepts S, then A accepts $t_1 S = t_1 s_1 s_2 \ldots s_m$. Hence, $t_1 S$ is a subsequence of T and this implies that S is a subsequence of T'.

□

Algorithm:
Input: DASG for the string $T = t_1 t_2 \ldots t_n$
Output: DASG for the string $T' = t_2 \ldots t_n$

1: **for** each $a \in \Sigma$ such that $\delta(q_0, a) \neq \emptyset$ **do**
2:    delete the transition $\delta(q_0, a)$
3: **end for**
4: $Q \leftarrow Q \setminus \{q_0\}$

Time complexity: the algorithm requires $\mathcal{O}(|\Sigma|)$ time.

4   Adding a State on the Right

The operation is required if we lengthen string T by a symbol x. Then, $T' = Tx = t_1 t_2 \ldots t_n x$. The DASG for the string T' is $A' = (Q', \Sigma, \delta', q_0, Q')$, where $Q' = Q \cup \{q\}$ and δ' is defined as follows:
$\delta'(p, a) = \delta(p, a)$ for all $p \in Q$ and all $a \in \Sigma$,
$\delta'(p, x) = q$ for all $p \in Q$ such that $\delta(p, x) = \emptyset$.

Lemma 3 A' accepts a string $S = s_1 s_2 \ldots s_m$ if and only if S is a subsequence of the string T'.

Proof. We prove the Lemma in two steps.

1. A' accepts all subsequences of T': For any subsequence S of T' the following holds: $S = R$, or $S = Rx$, where R is a subsequence of T. Let us consider both cases:
   (a) $S = R$: A accepts S and therefore A' accepts S as well.
   (b) $S = Rx$: A accepts R, that is, there exists a state $p \in Q$, such that $\delta^*(q_0, R) = p$. According to definition the same transitions also exist in A' and p has a transition for the symbol x in A'. Hence, A' accepts S.
2. If S is accepted by A', then S is a subsequence of T': There are two possibilities:
   (a) $s_m = x$: $s_1 s_2 \ldots s_{m-1}$ is accepted by A, that is, $s_1 s_2 \ldots s_{m-1}$ is a subsequence of T. Consequently, S is a subsequence of T'.
   (b) $s_m \neq x$: S is accepted by A, that is, S is a subsequence of T and hence S is a subsequence of T' as well.

$\square$

Algorithm:
Input: DASG for the string $T = t_1 t_2 \ldots t_n$
Output: DASG for the string $T' = t_1 t_2 \ldots t_n x$

1: $i \leftarrow n$
2: **while** $i \geq 0$ and $\delta(q_i, x) = \emptyset$ **do**
3:    $\delta(q_i, x) \leftarrow q$
4:    $i \leftarrow i - 1$
5: **end while**
6: $Q \leftarrow Q \cup \{q\}$

Time complexity: the algorithm requires $\mathcal{O}(n)$ time in the worst case.

5 Deleting a State on the Right

The operation is required if we delete the last symbol of the string T. Then, $T^{\iota} = t_1 t_2 \ldots t_{n-1}$. The DASG for the string T^{ι} is $A^{\iota} = (Q^{\iota}, \Sigma, \delta^{\iota}, q_0, Q^{\iota})$, where $Q^{\iota} = Q \setminus \{q_n\}$ and δ^{ι} is defined as follows:

$\delta^{\iota}(p, a) = \delta(p, a)$ for all $p \in Q$ and all $a \in \Sigma, a \neq x$,

$\delta^{\iota}(p, x) = \emptyset$ for all $p \in Q$ such that $\delta(p, x) = q_n$,

$\delta^{\iota}(p, x) = \delta(p, x)$ for all $p \in Q$ such that $\delta(p, x) \neq q_n$.

Lemma 4 A^{ι} *accepts a string* $S = s_1 s_2 \ldots s_m$ *if and only if* S *is a subsequence of the string* T^{ι}.

Proof. We prove the Lemma in two steps.

1. A^{ι} accepts all subsequences of T^{ι}: If S is a subsequence of T^{ι}, then $St_n = s_1 s_2 \ldots s_m t_n$ is a subsequence T and St_n is accepted by A. Hence, $\delta^*(q_0, S) = q_k, 0 \leq k < n$. According to definition the same transitions exist in A^{ι} and the following holds: $\delta^{\iota *}(q_0, S) = q_k$. Hence, S is accepted by A^{ι}.

2. If S is accepted by A^{ι}, then S is a subsequence of T^{ι}: If S is accepted by A^{ι}, then $St_n = s_1 s_2 \ldots s_m t_n$ is accepted by A, that is, St_n is a subsequence of T. Consequently, S is a subsequence of T^{ι}.   □

Algorithm:
Input: DASG for the string $T = t_1 t_2 \ldots t_n$
Output: DASG for the string $T^{\iota} = t_1 t_2 \ldots t_{n-1}$

1: $i \leftarrow n - 1$
2: **while** $i \geq 0$ and $\delta(q_i, t_i) = q_n$ **do**
3:     $\delta(q_i, t_i) \leftarrow \emptyset$
4:     $i \leftarrow i - 1$
5: **end while**
6: $Q \leftarrow Q \setminus \{q_n\}$

Time complexity: the algorithm requires $\mathcal{O}(n)$ time in the worst case.

6 Adding an Inner State

The operation is required if we insert a symbol x into the string T. Then, the modified string $T^{\iota} = t_1 \ldots t_i x t_{i+1} \ldots t_n$. The DASG for the string T^{ι} is $A^{\iota} = (Q^{\iota}, \Sigma, \delta^{\iota}, q_0, Q^{\iota})$, where $Q^{\iota} = Q \cup \{q\}$ and δ^{ι} is defined as follows:

$\delta^{\iota}(p, a) = \delta(p, a)$ for all $p \in Q$ and all $a \in \Sigma, a \neq x$,

$\delta^{\iota}(q, a) = \delta(q_i, a)$ for all $a \in \Sigma$,

$\delta^{\iota}(p, x) = \delta(p, x)$ for all $p \in \{q_{i+1}, \ldots, q_n\}$,

for all $p \in \{q_0, \ldots, q_i\}$: $\delta^{\iota}(p, x) = q$ if $\delta(p, x) = \delta(q_i, x)$, $\delta^{\iota}(p, x) = \delta(p, x)$ otherwise.

Lemma 5 *A^{ι} accepts a string $S = s_1 s_2 \ldots s_m$ if and only if S is a subsequence of the string T^{ι}.*

Proof. Let $T_1 = t_1 \ldots t_i$ and $T_2 = t_{i+1} \ldots t_n$. We prove the Lemma in two steps.

1. A^{ι} accepts all subsequences of T^{ι}: Let us consider maximal j such that $s_1 s_2 \ldots s_j$ is a subsequence of T_1 and denote $S_1 = s_1 s_2 \ldots s_j$. Then, $s_{j+1} \neq x$, or $s_{j+1} = x$.

   (a) $s_{j+1} \neq x$: As $S_1 T_2$ is a subsequence of T, it is ¿accepted by A. Therefore the following holds: $\delta^*(q_0, S_1) = q_k, 0 \leq k \leq i$. As $s_{j+1} \neq x$, $\delta(q_k, s_{j+1}) = q_l, (i+1) \leq l \leq n$ and $\delta^*(q_l, s_{j+2} \ldots s_m) = q_p, l \leq p \leq n$. According to definition the same transitions exist in A^{ι}. Hence, A^{ι} accepts S.

   (b) $s_{j+1} = x$: $\delta^*(q_0, S_1) = q_k, 0 \leq k \leq i$ and according to definition the same transitions exist in A^{ι}: $\delta^{\iota*}(q_0, S_1) = q_k$. Then, $\delta^{\iota}(q_k, x) = q$ and according to definition q in A^{ι} has the same transitions as q_i in A: $\delta(q_i, s_{j+1}) = \delta^{\iota}(q, s_{j+1})$. As $s_{j+1} \ldots s_m$ is a subsequence of T_2, A accepts $t_1 \ldots t_i s_{j+1} \ldots s_m$. Hence the following holds: $\delta^*(q_i, s_{j+1} \ldots s_m) = q_l, i \leq l \leq n$ and according to definition also $\delta^{\iota*}(q, s_{j+1} \ldots s_m) = q_l$. Consequently, A^{ι} accepts S.

2. If S is accepted by A^{ι}, then S is a subsequence of T^{ι}: Let us consider maximum j such that $\delta^{\iota*}(q_0, s_1 \ldots s_j) = q_k, 0 \leq k \leq i$ and denote $S_1 = s_1 s_2 \ldots s_j$. Then, $s_{j+1} \neq x$, or $s_{j+1} = x$.

   (a) $s_{j+1} \neq x$: $\delta^{\iota}(q_k, s_{j+1}) = \delta(q_k, s_{j+1}) = q_l, where (i+1) \leq l \leq n$, and also $\delta^{\iota*}(q_l, s_{j+2} \ldots s_m) = \delta^*(q_l, s_{j+2} \ldots s_m) = q_p, l \leq p \leq n$. Hence, A accepts S and S is a subsequence of T. Any subsequence of T is a subsequence of T^{ι} as well.

   (b) $s_{j+1} = x$: $\delta^{\iota}(q_k, x) = q$ and according to definition q in A^{ι} has the same transitions as q_i in A. Consequently, the following also holds: $\delta^{\iota*}(q, s_{j+2} \ldots s_m) = \delta^*(q_i, s_{j+2} \ldots s_m) = q_p, where i \leq p \leq n$. Hence, $S_1 t_{i+1} \ldots t_n$ and $t_1 \ldots t_i s_{j+2} \ldots s_m$ are accepted by A. Then, S_1 is a subsequence of T_1 and $s_{j+2} \ldots s_m$ is a subsequence of T_2. Consequently $S = S_1 x s_{j+2} \ldots s_m$ is a subsequence of $T^{\iota} = T_1 x T_2$. □

Algorithm:
Input: DASG for the string $T = t_1 t_2 \ldots t_n$
Output: DASG for the string $T^{\iota} = t_1 \ldots t_i x t_{i+1} \ldots t_n$

```
1: for each a ∈ Σ do
2:    δ(q, a) ← δ(q_i, a)
3: end for
4: old ← δ(q_i, x)
5: j ← i
6: while j ≥ 0 and δ(q_j, x) = old do
7:    δ(q_j, x) ← q
8:    j ← j − 1
9: end while
10: Q ← Q ∪ {q}
```

Time complexity: the algorithm requires $\mathcal{O}(n)$ time in the worst case.

7 Deleting an Inner State

The operation is required if we delete an inner symbol t_i from the string T. Then, $T' = t_1 \ldots t_{i-1} t_{i+1} \ldots t_n$. The DASG for the string T' is $A' = (Q', \Sigma, \delta', q_0, Q')$, where $Q' = Q \setminus \{q_i\}$ and δ' is defined as follows:

$\delta'(p, a) = \delta(p, a)$ for all $p \in Q'$ and all $a \in \Sigma, a \neq t_i$,

$\delta'(p, t_i) = \delta(p, t_i)$ for all $p \in \{q_{i+1}, \ldots, q_n\}$,

for all $p \in \{q_0, \ldots, q_{i-1}\}$: $\delta'(p, t_i) = \delta(q_i, t_i)$ if $\delta(p, t_i) = q_i$, $\delta'(p, t_i) = \delta(p, t_i)$ otherwise.

Lemma 6 A' accepts a string $S = s_1 s_2 \ldots s_m$ if and only if S is a subsequence of the string T'.

Proof. Let $T_1 = t_1 \ldots t_{i-1}$ and $T_2 = t_{i+1} \ldots t_n$. We prove the Lemma in two steps.

1. A' accepts all subsequences of T': Let us consider maximal j such that $s_1 s_2 \ldots s_j$ is a subsequence of T_1 and denote $S_1 = s_1 s_2 \ldots s_j$. Then, $s_{j+1} \neq t_i$, or $s_{j+1} = t_i$.

   (a) $s_{j+1} \neq t_i$: As $S_1 T_2$ is a subsequence of T, it is accepted by A. Therefore the following holds: $\delta^*(q_0, S_1) = q_k, 0 \leq k < i$. As $s_{j+1} \neq t_i, \delta(q_k, s_{j+1}) = q_l, (i+1) \leq l \leq n$ and $\delta^*(q_l, s_{j+2} \ldots s_m) = q_r, l \leq r \leq n$. According to definition the same transitions exist in A'. Hence, A' accepts S.

   (b) $s_{j+1} = t_i$: $\delta^*(q_0, S_1) = q_k, 0 \leq k \leq i$ and according to definition the same transitions exist in A': $\delta'^*(q_0, S_1) = q_k$. Also the following holds: $\delta(q_k, t_i) = q_i$) and according to definition $\delta'(q_k, t_i) = \delta(q_i, t_i) = q_l, k \leq l \leq n$. As $S_1 t_i s_{j+1} \ldots s_m$ is a subsequence of T, it is accepted by A and therefore the following holds: $\delta^*(q_l, s_{j+1} \ldots s_m) = q_r, l \leq r \leq n$ and according to definition also $\delta'^*(q_l, s_{j+1} \ldots s_m) = q_r$. Hence, A' accepts $S_1 s_{j+1} \ldots s_m$.

2. If S is accepted by A', then S is a subsequence of T': Let us consider maximal j such that $\delta'^*(q_0, s_1 \ldots s_j) = q_k, 0 \leq k \leq (i-1)$ and denote $S_1 = s_1 \ldots s_j$. According to definition: $\delta^*(q_0, S_1) = q_k$. Then, $s_{j+1} \neq t_i$, or $s_{j+1} = t_i$.

   (a) $s_{j+1} \neq t_i$: $\delta'(q_k, s_{j+1}) = \delta(q_k, s_{j+1}) = q_l, (i+1) \leq l \leq n$. As A accepts $S_1 t_i \ldots t_n$ and $t_1 \ldots t_i s_{j+1} \ldots s_m$, S_1 is a subsequence of T_1 and $s_{j+1} \ldots s_m$ is a subsequence of T_2. Hence, $S_1 s_{j+1} \ldots s_m$ is a subsequence of T'.

   (b) $s_{j+1} = t_i$: $\delta'(q_k, t_i) = \delta(q_i, t_i) = q_l, (i+1) \leq l \leq n$. Again, as A accepts $S_1 t_i \ldots t_n$ and $t_1 \ldots t_i s_{j+1} \ldots s_m$, S_1 is a subsequence of T_1 and $s_{j+1} \ldots s_m$ is a subsequence of T_2. Hence, $S_1 s_{j+1} \ldots s_m$ is a subsequence of T'.

□

Algorithm:
Input: DASG for the string $T = t_1 t_2 \ldots t_n$
Output: DASG for the string $T' = t_1 \ldots t_{i-1} t_{i+1} \ldots t_n$

```
 1: new ← δ(qi, ti)
 2: j ← i - 1
 3: while j ≥ 0 and δ(qj, ti) = qi do
 4:    δ(qj, ti) ← new
 5:    j ← j - 1
 6: end while
 7: for each a ∈ Σ such that δ(qi, a) ≠ ∅ do
 8:    delete the transition δ(qi, a)
 9: end for
10: Q ← Q \ {qi}
```

Time complexity: the algorithm requires $\mathcal{O}(n + |\Sigma|)$ time in the worst case.

8   Replacing a Transition Label

The operation is required if we replace a symbol t_i of the string T by the symbol x. Then, $T^{\iota} = t_1 \ldots t_{i-1} x t_{i+1} \ldots t_n$. If $x \notin \Sigma$, then we define $\delta(q, x) = \emptyset$ for all $q \in Q$. The DASG for the string T^{ι} is $A^{\iota} = (Q, \Sigma \cup \{x\}, \delta^{\iota}, q_0, Q)$, where δ^{ι} is defined as follows:

$\delta^{\iota}(p, a) = \delta(p, a)$ for all $p \in Q$ and all $a \in \Sigma, a \neq x, a \neq t_i$,
$\delta^{\iota}(p, x) = \delta(p, x)$ for all $p \in \{q_i, \ldots, q_n\}$,
$\delta^{\iota}(p, t_i) = \delta(q_i, t_i)$ for all $p \in Q$ such that $\delta(p, t_i) = q_i$,
$\delta^{\iota}(p, t_i) = \delta(p, t_i)$ for all $p \in Q$ such that $\delta(p, t_i) \neq q_i$,
for all $p \in \{q_0, \ldots, q_{i-1}\}$: $\delta^{\iota}(p, x) = q_i$ if $\delta(p, x) = \delta(q_i, x)$, $\delta^{\iota}(p, x) = \delta(p, x)$ otherwise.

Lemma 7 A^{ι} accepts a string $S = s_1 s_2 \ldots s_m$ if and only if S is a subsequence of the string T^{ι}.

Proof. Let $T_1 = t_1 \ldots t_{i-1}$ and $T_2 = t_{i+1} \ldots t_n$. We prove the Lemma in two steps.

1. A^{ι} accepts all subsequences of T^{ι}: Let us consider maximal j such that $s_1 s_2 \ldots s_j$ is a subsequence of T_1 and denote $S_1 = s_1 s_2 \ldots s_j$. Then, $s_{j+1} \neq x$, or $s_{j+1} = x$.

   (a) $s_{j+1} \neq x$: As $S_1 t_i \ldots t_n$ is a subsequence of T, it is accepted by A and the following holds: $\delta^*(q_0, S_1) = q_k, 0 \leq k \leq (i-1)$. According to definition the same transitions exist in A^{ι}: $\delta^{\iota*}(q_0, S_1) = q_k$. Again, according to definition: $\delta^{\iota}(q_k, s_{j+1}) = \delta(q_k, s_{j+1}) = q_l, (i+1) \leq l \leq n$ and $\delta^*(q_l, s_{j+2} \ldots s_m) = \delta^{\iota*}(q_l, s_{j+2} \ldots s_m) = q_r, l \leq r \leq n$. Consequently, A^{ι} accepts S.

   (b) $s_{j+1} = x$: As $S_1 t_i \ldots t_n$ is a subsequence of T, the following holds: $\delta^*(q_0, S_1) = q_k, 0 \leq k \leq (i-1)$ and according to definition the following holds: $\delta^{\iota}(q_k, x) = q_i$. As $s_{j+2} \ldots s_m$ is a subsequence of T_2, A accepts $t_1 \ldots t_i s_{j+2} \ldots s_m$: $\delta^*(q_i, s_{j+2} \ldots s_m) = q_r, i \leq r \leq n$. According to definition $\delta^{\iota*}(q_i, s_{j+2} \ldots s_m) = q_r$. Consequently, A^{ι} accepts S.

2. If S is accepted by A', then S is a subsequence of T': Let us consider maximal j such that $\delta'^*(q_0, s_1 \dots s_j) = q_k, 0 \le k \le (i-1)$ and denote $S_1 = s_1 \dots s_j$. According to definition: $\delta^*(q_0, S_1) = q_k$. Then, $s_{j+1} \ne x$, or $s_{j+1} = x$.

   (a) $s_{j+1} \ne x$: $\delta'(q_k, s_{j+1}) = q_l, (i+1) \le l \le n$ and according to definition $\delta'(q_k, s_{j+1}) = q_l$. Consequently, $S_1 t_i \dots t_n$ and $t_1 \dots t_i s_{j+1} \dots s_m$ are both accepted by A. This implies that S_1 is a subsequence of T_1 and $s_{j+1} \dots s_m$ is a subsequence of T_2. Hence, $S = S_1 s_{j+1} \dots s_m$ is a subsequence of T'.

   (b) $s_{j+1} = x$: According to definition: $\delta(q_k, x) = q_i$ and $\delta'^*(q_i, s_{j+2} \dots s_m) = \delta^*(q_i, s_{j+2} \dots s_m)$. Consequently, $S_1 t_i \dots t_n$ and $t_1 \dots t_i s_{j+2} \dots s_m$ are both accepted by A. This implies that S_1 is a subsequence of T_1 and $s_{j+2} \dots s_m$ is a subsequence of T_2. Hence, $S = S_1 x s_{j+2} \dots s_m$ is a subsequence of T'.

$\square$

Algorithm:
Input: DASG for the string $T = t_1 t_2 \dots t_n$
Output: DASG for the string $T' = t_1 \dots t_{i-1} x t_{i+1} \dots t_n$

```
1: j ← i − 1
2: while j ≥ 0 and δ(q_j, t_i) = q_i do
3:     δ(q_j, t_i) ← δ(q_i, t_i)
4: end while
5: j ← i − 1
6: while j ≥ 0 and δ(q_j, x) = δ(q_i, x) do
7:     δ(q_j, x) ← q_i
8: end while
```

Time complexity: the algorithm requires $\mathcal{O}(n)$ time in the worst case.

9 Conclusion

We have defined the following operations on DASG: adding a state on the left, deleting a state on the left, adding a state on the right, deleting a state on the right, adding an inner state, deleting an inner state, and replacing a transition label. For each of these operations, we have provided the modification of DASG and the proof of its correctness.

Note: Dominique Revuz suggested the representation of DASG, in which each state has lists of states for each letter of the alphabet. Each state has in a list the states that have a transition to this state (for each letter of the alphabet). With this representation are some operations on DASG much simpler.

References

[1] Baeza-Yates, R. A.: Searching subsequences. Theoretical Computer Science 78 (1991), pages 363–378.

[2] Crochemore, M., Rytter, W.: Text algorithms. Oxford University Press, 1994.

[3] Hirschberg, D. S.: A linear space algorithm for computing maximal common sub-sequences. Communication of ACM, 18(6), 1975, pages 341–343.

[4] Myers, E. W., Miller, W.: Optimal alignments in linear space. Computer Applications in the Biosciences, 4(1), 1988, pages 11– 17.

[5] Sankoff, D., Kruskal, J. B.: Time warps, string edits, and macromolecules: the theory and practice of sequence comparison. Addison-Wesley, Reading, MA, 1983.

[6] Tronìček, Z., Melichar, B.: Directed acyclic subsequence graph. Prague Stringology Club Workshop, 1998.

Implementation of Nondeterministic Finite Automata for Approximate Pattern Matching[*]

Jan Holub and Bořivoj Melichar

Department of Computer Science and Engineering
Czech Technical University
Karlovo nám. 13, 121 35 Prague 2, Czech Republic
{holub,melichar}@cs.felk.cvut.cz

Abstract. There are two ways of using the nondeterministic finite automata (*NFA*). The first one is the transformation to the equivalent deterministic finite automaton and the second one is the simulation of the run of *NFA*. In this paper we discuss the second way. We present an overview of the simulation methods that have been found in the approximate string matching. We generalize these simulation methods and form the rules for the usage of these methods.

1 Introduction

The nondeterministic finite automaton (*NFA*) is a quintuple $M = (Q, \Sigma, \delta, I, F)$ where Q is a finite set of states, Σ is a finite input alphabet, δ is a mapping from $Q \times (\Sigma \cup \{\varepsilon\})$ to set $\mathcal{P}(Q)$ of all the subsets of Q, $I \subseteq Q$ is a set of the initial states and F is a set of the final states.

NFA cannot be directly used because of its nondeterminism — there are some states $q \in Q$ from which *NFA* can move for an input symbol $a \in \Sigma$ to more than one state because it may be $|\delta(q, a)| > 1$. There are two possibilities of using *NFA*:

1. to transform *NFA* to the equivalent deterministic finite automaton (*DFA*),
2. to simulate the run of *NFA* in deterministic way.

DFA has only one initial state, $\delta(q, \varepsilon) = \{q\}$, $\forall q \in Q$, and $|\delta(q, a)| \leq 1$, $\forall q \in Q$, $\forall a \in \Sigma$. If *DFA* has to perform a transition $\delta(q, a)$, $q \in Q$, $a \in \Sigma$, such that $\delta(q, a) = \emptyset$, the computation ends without reaching any final state. In some definitions [HU79] there is required $|\delta(q, a)| = 1$, $\forall q \in Q$, $\forall a \in \Sigma$. To fulfil this condition we can insert new state q' and modify δ such that $\delta(q', a) = \{q'\}$, $\forall a \in \Sigma$, and each transition $\delta(q, a) = \emptyset$, $q \in Q$, $a \in \Sigma$, we replace by transition $\delta(q, a) = \{q'\}$. But in this case the *DFA* has to process whole input string to get the result of the computation even if it is known before the end of the input string (*DFA* is in the state that cannot reach any final state).

[*] This research was partially supported by grant 201/98/1155 of the Grant Agency of Czech Republic and by internal grant 3098098/336 of Czech Technical University.

The transformation of *NFA* to the equivalent *DFA* using the standard subset construction eliminating inaccessible states may lead to *DFA* with 2^m states and take time $\mathcal{O}(2^m)$, where m is a number of states of the *NFA*. The transformation is shown in Figure 1 where $\varepsilon CLOSURE(P)$ is a set $\{q' \mid q' \in \delta(q, \varepsilon), q \in P\} \cup P$.

After the transformation *DFA* runs in time $\mathcal{O}(n)$ where n is the length of the input string.

Algorithm
Input: *NFA* $M = (Q, \Sigma, \delta, I, F)$.
Output: *DFA* M' accepting language $L(M)$.
Method: $M' = (Q', \Sigma, \delta', q'_0, F')$ where Q', δ', q'_0, and F' are constructed in the following way:

> $q'_0 := \varepsilon CLOSURE(I)$
> $Q' := \emptyset$
> $S := \{q'_0\}$ /* the set of not yet processed states */
> **for each** $q' \in S$ **do**
>   **for each** $a \in \Sigma$ **do**
> $\delta'(q', a) := \bigcup_{q \in q'} \varepsilon CLOSURE(\delta(q, a))$
>   **if** $\delta'(q', a) \notin Q'$ **and** $\delta'(q', a) \notin S$ **then**
> $S := S \cup \delta'(q', a)$
>   **endif**
>   **endfor**
>   **if** $q' \cap F \neq \emptyset$ **then**
> $F' := F' \cup \{q'\}$
>   **endif**
> $S := S - \{q'\}$
> $Q' := Q' \cup \{q'\}$
> **endfor**

Fig. 1. The transformation of *NFA* to the equivalent *DFA* without inaccessible states.

On the other hand the simulation of the run of *NFA* has higher the running time but we can save some space and the preprocessing time. In case that the space complexity of *DFA* makes *DFA* unusable, we have to use some of simulation methods.

2   Naive Method

In the simulation of the run of *NFA* the set of active states is held. At the beginning of the simulation the set of active states contains all the initial states. Then in each step of the simulation a new set of active states is computed by the evaluation of the transitions for all active states of the previous step. The naive method is very similar to the transformation of *NFA* to the equivalent

DFA using the standard subset construction eliminating inaccessible states and is shown in Figure 2.

Algorithm
Input: *NFA* $M = (Q, \Sigma, \delta, I, F)$, input text $T = t_1 t_2 \ldots t_n$.
Output: Output of run of *NFA*.
Method: Set S of active states is used.

   $S := I$
   $i := 1$
   while $i \leq n$ **and** $S \neq \emptyset$ **do**
      $S := \bigcup_{q \in S} \varepsilon CLOSURE(\delta(q, t_i))$
      if $S \cap F \neq \emptyset$ **then**
         write(information associated with each final state in S)
      endif
      $i := i + 1$
   endwhile

Fig. 2. The simulation of the run of *NFA* — the naive method.

This naive simulation runs in time $\mathcal{O}(|Q|n)$ and space $\mathcal{O}(|Q||\Sigma|)$ where $|Q|$ is the number of states of *NFA* and n is the length of the input string. All other simulation methods are based on this naive method and they differ only in the implementation of the set of active states.

We can directly implement the naive method such that the set of active states would be implemented by the bit vector (0 represents that the corresponding state is active and 1 represents that the state is not active; the meaning of 0 and 1 can be also exchanged). This method is shown in Figure 3. Operators **AND** and **OR** are bitwise operation AND and OR, respectively. In this case it is required *NFA* without ε-transitions. This condition does not restrict the simulation since each *NFA* with ε-transitions can be transformed to equivalent *NFA* without ε-transitions. In Figure 3 we can modify the test, whether a final state has been reached, to the test, which final state has been reached. In such case we can report the information associated with such final state.

We have found out two simulation methods called the *dynamic programming* and the *bit parallelism*.

3   Dynamic Programming

In the dynamic programming we divide the set Q of states of *NFA* to subsets $Q_1, Q_2, \ldots, Q_l$ such that:

(1) $Q = \bigcup_{i=1}^{l} Q_i$,
(2) $Q_i \cap Q_j = \emptyset$, $\forall i, j, 1 \leq i \leq l, 1 \leq j \leq l, i \neq j$,
(3) there can be at most one active state in each subset Q_i, $1 \leq i \leq l$.

Algorithm
Input: Number $|Q|$ of states of *NFA*, transition table C of size $|Q| \times |\Sigma|$ represented by the bit vectors, bit vector I of the initial states, bit vector F of the final states, and input text $T = t_1 t_2 \ldots t_n$.
Output: Output of the run of *NFA*.
Method: Vector S_i of active states is used.

$S_0 := I$
$i := 1$
while $i \leq n$ **and** $S_{i-1} \neq (1, 1, \ldots, 1)$ **do**
$\quad S_i := (1, 1, \ldots, 1)$
$\quad$ **for** $j := 1, 2, \ldots, |Q|$ **do**
$\quad\quad$ **if** $S_{i-1,j} = 0$ **then** $S_{i,j} := S_{i,j}$ AND c_{j,t_i}
$\quad$ **endfor**
$\quad$ **if** $(S_i$ OR $F) \neq (1, 1, \ldots, 1)$ **then**
$\quad\quad$ **write**('*NFA* has reached a final state')
$\quad$ **endif**
$\quad i := i + 1$
endwhile

Fig. 3. The simulation of the run of *NFA* — the naive method implemented by the bit vectors.

The states in each subset are numbered. Each subset Q_i is implemented by the integer variable that contains the number of active state in this subset or a special value if there is no active state in this subset.

In the dynamic programming our goal is to minimize the number of integer variables (number of subsets) and to implement the transitions in the simplest way.

This method has been used in the approximate string matching using the Levenshtein distance which is defined as a searching of all the occurrences of pattern $P = p_1 p_2 \ldots p_m$ in text $T = t_1 t_2 \ldots t_n$ such that the pattern can be converted to the found string using at most k edit operations *replace* (one character is replaced by another), *insert* (one symbol is inserted), and *delete* (one symbol is deleted).

The algorithm has been presented in [Sel80, Ukk85] but not as a simulation of the run of the *NFA*. *NFA* for the approximate string matching, shown in Figure 4[1], has been presented in [Mel96, Hol96] and it has $m+1$ depths (columns) and $k + 1$ levels (rows) — each level of states for each edit distance less or equal to k. The way how the dynamic programming simulates the run of the *NFA* is shown in [Mel96, Hol98a].

In the dynamic programming there is for each depth of *NFA* one integer variable that contains the lowest number of level of active state on this level or value greater than or equal to $k + 1$ if there is no active state in this depth. The

---

[1] Symbol $\bar{p}$ in the figure represents $\Sigma - \{p\}$.

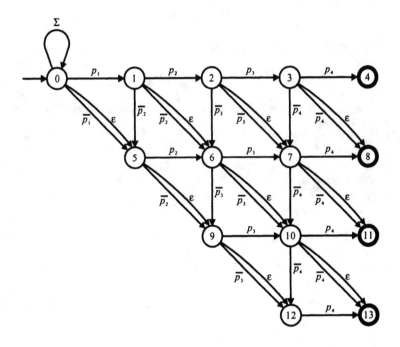

Fig. 4. *NFA* for the approximate string matching using the Levenshtein distance ($m = 4$, $k = 2$).

formula for computing the vector D_i, $0 \le i \le n$, of the integer variables is as follows:

$$
\begin{aligned}
&d_{j,0} := j, && 0 \le j \le m \\
&d_{0,i} := 0, && 0 \le i \le n \\
&d_{j,i} := \textbf{if } t_i = p_j \textbf{ then } d_{j-1,i-1} && \\
&\quad\quad \textbf{else min}(d_{j-1,i-1}+1, d_{j,i-1}+1, && \\
&\quad\quad\quad\quad d_{j-1,i}+1) && 0 < i \le n, 0 < j \le m
\end{aligned}
\tag{1}
$$

The dynamic programming algorithm [Sel80, Ukk85] has been designed to compute the edit distances of the prefixes of pattern P — the Levenshtein edit distance between the string ending at position i in text T and the prefix of length j of pattern P is $d_{j,i}$. In this case the reality is described by the vector of the edit distances that also describes the active states of *NFA*.

This simulation runs in time $\mathcal{O}(mn)$ and needs space $\mathcal{O}(m)$. There are also some optimizations of [Sel80, Ukk85] that differ only in the implementation of the matrix D, ie. [GP89] with time $\mathcal{O}(kn)$.

In [Hol98b] we have designed another simulation based on the dynamic programming. We have for all the states located on the same diagonal one integer variable because if there is an active state in the diagonal, then all the states located lower on the same diagonal are also active due to the ε-transitions. Therefore for each diagonal we store the level of the highest active state on the diagonal. This simulation technique is used in case that we are interested in all the occurrences of the pattern with at most k errors but we do not want to know the number of errors in the found strings. In this case we can remove states located on the diagonal that have less than $k + 1$ states because these states are needed only to determine the number of errors in the found string. This simulation runs in time $\mathcal{O}((m - k)n)$.

4   Bit Parallelism

The bit parallelism is very similar to the naive method implemented by the bit vectors but in this method we do not need the transition table. The set Q of states of NFA is divided to subsets $Q_1, Q_2, \ldots, Q_l$ such that:

(1) $Q = \bigcup_{i=1}^{l} Q_i$,
(2) $Q_i \cap Q_j = \emptyset$, $\forall i, j,\ 1 \leq i \leq l,\ 1 \leq j \leq l,\ i \neq j$.

Each such subset Q_i is implemented by the bit vector in which each bit represents state $q \in Q_i$ — 0 represents that state q is active. The transition table in this technique is represented by the formula for the computation of the vectors in which groups of transitions are computed at once (in parallel — therefore this technique is called the bit parallelism).

As an example of this technique we show Shift-Or algorithm [BYG92] for the approximate string matching using the Levenshtein distance (there are also the modifications Shift-Add [BYG92] and Shift-Add [WM92]). In [Hol96, Hol97] we have shown that the Shift-Or algorithm simulates the run of NFA for the approximate string matching shown in Figure 4.

In this algorithm there is for each level j, $0 \leq j \leq k$, of the states of NFA one bit vector R^j. To implement the horizontal transitions the whole vector is shifted to the right by using the bitwise operation **shl**[2] which inserts 0 at the beginning of the vector in order to implement the self-loop of the initial state. Then the mask vector is used in order to select only such transitions that correspond to the input symbol. It is performed by bitwise operation **AND** which sets the active states in the positions not corresponding to the input symbol to the inactive states.

The implementation of the diagonal transitions is performed by adding the shifted vector of level j, $0 \leq j < k$, to the new value of the vector of level $j + 1$. This adding it performed by bitwise operation **OR** which inserts the active states (represented by 0) to the positions on which there are the inactive states. The ε-transitions and the vertical transitions are implemented in the similar way.

---

[2] In the implementation of Shift-Or algorithm the vectors are shifted to the left that is more suitable for the case that the vector is longer than the used computer word and it should be divided into two or more computer words.

The formula for computing the vectors is as follows:

$$
\begin{aligned}
r^l_{j,0} &:= 0, & 0 < j \le l, 0 < l \le k \\
r^l_{j,0} &:= 1, & l < j \le m, 0 \le l \le k \\
R^0_i &:= \mathbf{shl}(R^0_{i-1}) \text{ OR } D[t_i], & 0 < i \le n \\
R^l_i &:= (\mathbf{shl}(R^l_{i-1}) \text{ OR } D[t_i]) \\
&\quad \text{AND } \mathbf{shl}(R^{l-1}_{i-1} \text{ AND } R^{l-1}_i) \text{ AND } R^{l-1}_{i-1}, 0 < i \le n, 0 < l \le k
\end{aligned}
\tag{2}
$$

Where each element $d_{j,x}$, $0 < j \le m$, $x \in \Sigma$, contains 0, if $p_j = x$, or 1, otherwise.

This simulation technique was also used for *NFA* s for the approximate sequence matching [Hol97] using the Hamming, Levenshtein, and generalized Levenshtein distances.

5   Conclusion

We have presented two simulation techniques for *NFA*. They can be used in the case that the transformation of the *NFA* to the equivalent *DFA* cannot be used because of its space complexity. It can be also used in the case that the input text is not so long and the time needed for the transformation of *NFA* to the equivalent *DFA* and the time for the run of *DFA* together is greater than the time of some simulation of the *NFA*.

References

[BYG92]   R. A. Baeza-Yates and G. H. Gonnet. A new approach to text searching. *Commun. ACM*, 35(10):74–82, 1992.

[GP89]   Z. Galil and K. Park. An improved algorithm for approximate string matching. In G. Ausiello, M. Dezani-Ciancaglini, and S. Ronchi Della Rocca, editors, *Proceedings of the 16th International Colloquium on Automata, Languages and Programming*, number 372 in Lecture Notes in Computer Science, pages 394–404, Stresa, Italy, 1989. Springer-Verlag, Berlin.

[Hol96]   J. Holub. Reduced nondeterministic finite automata for approximate string matching. In J. Holub, editor, *Proceedings of the Prague Stringologic Club Workshop '96*, pages 19–27, Prague, Czech Republic, 1996. Collaborative Report DC-96-10.

[Hol97]   J. Holub. Simulation of NFA in approximate string and sequence matching. In J. Holub, editor, *Proceedings of the Prague Stringology Club Workshop '97*, pages 39–46, Czech Technical University, Prague, Czech Republic, 1997. Collaborative Report DC–97–03.

[Hol98a]   J. Holub. Simulation of nondeterministic finite automata in approximate string matching. In *WORKSHOP '98*, pages 211–212, Czech Technical University, Prague, Czech Republic, February 1998.

[Hol98b]   J. Holub. Simulation of nondeterministic finite automata in approximate string and sequence matching. Technical report, Czech Technical University, Prague, Czech Republic, April 1998. Research Report DC–98–04.

[HU79]   J. E. Hopcroft and J. D. Ullman. *Introduction to automata, languages and computations.* Addison-Wesley, Reading, MA, 1979.

[Mel96]   B. Melichar. String matching with k differences by finite automata. In *Proceedings of the 13th International Conference on Pattern Recognition*, volume II., pages 256–260, Vienna, Austria, 1996. IEEE Computer Society Press.

[Sel80]   P. H. Sellers. The theory and computation of evolutionary distances: Pattern recognition. *J. Algorithms*, 1(4):359–373, 1980.

[Ukk85]   E. Ukkonen. Finding approximate patterns in strings. *J. Algorithms*, 6(1–3):132–137, 1985.

[WM92]   S. Wu and U. Manber. Fast text searching allowing errors. *Commun. ACM*, 35(10):83–91, 1992.

The Syntactic Prediction with Token Automata: Application to HandiAS System

Denis Maurel[1], Brigitte Le Pévédic[2], and Olivier Rousseau[3]

[1] LI/E3i
Université de Tours
64 avenue Jean-Portalis, F37200 Tours, France
maurel@univ-tours.fr
[2] IRIN
Université de Nantes
2 rue de la Houssinière, F44072 Nantes cedex 03, France
brigitte.lepevedic@irin.univ-nantes.fr
[3] LI/E3i
and
C Technologie
2 rue Marie-Curie, F44470 Carquefou, France

Abstract. This paper presents an adjustable way of syntactic prediction on the basis of left context with token automata. Our purpose is to predict the end of a word on the basis of its first letters keyboarding. We illustrate our intention with the presentation of a prototype software for disabled communication aid, called *HandiAS*.

1 Motivation

This paper presents an adjustable way of syntactic prediction on the basis of left context with token automata. Our purpose is to predict the end of a word on the basis of its first letters keyboarding. Is it possible? This goal should be practically impossible, unless a syntactic study allows to guess its grammatical category and unless words are restricted to common vocabulary only...

We illustrate our intention with the presentation of a prototype software for disabled communication aid, called *HandiAS* [5]. It is an hybrid system, both symbolic and statistical. The symbolic part is based on the notion of sentence schema and acceptability, notions introduced by Z. S. Harris [3]. A sentence schema is easily represented by finite states automaton [10] [12] [7]. The statistical part is based on different studies about words token [1] [4] and on the notion of token schema that we are defining.

Figure 1 presents the interface of HandiAS: we are writing a text with a virtual keyboard and the system suggests a list of five words after each keyboarding.

First, we present a definition of a schema (§2); second, we describe our data base: the dictionary and the acceptability tables (§3); third, we explain how

J.-M. Champarnaud, D. Maurel, D. Ziadi (Eds.): WIA'98, LNCS 1660, pp. 100–109, 1999.

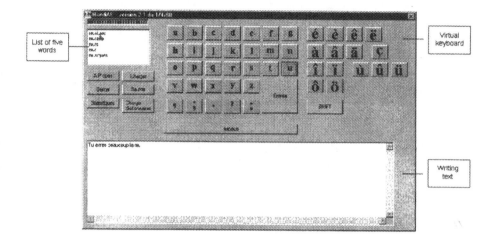

Fig. 1. The interface of HandiAS

the system works (§4), evolves and adapts to user (§5). We conclude with some results and prospects (§6).

2   Automata and Schemata

Definition 1 *The finite state machine $(Q, L, q_0, F, \delta, \lambda, \sigma)$ where:*

- *L is a set of grammatical categories*
- *(Q, L, q_0, F, δ) is a deterministic finite state automaton*
- *λ is a function: $Q \times L \to \mathbb{N}$ (transition token function)*
- *σ is a function: $F \to \mathbb{N}$ (final state token function)*

is name token schema[1]

The *HandiAS* system includes three schemata: we also drew our inspiration from the representation of F. Debili [2] who distinguishes some conjunctions, punctuations or prepositions as *break*. Therefore, our three schemata represent respectively sentences, noun phrases and verb phrases.

For instance, assume that one wants to write sentences such as: *Tu aimes beaucoup la musique.* (*You like music a lot.*), *Jean aime jouer de la musique.* (*John likes playing music.*), *Aimes-tu la musique?* (*Do you like music?*) or *Aimes-tu jouer de la musique?* (*Do you like playing music?*). The schema of Figure 2 makes it possible to recognize these four sentences. Of course, our true

---

[1] An token schema is not a string-to-weight transducer, as in [11]; we can't minimize it; it may not have convergent state[8]

schema is more complex: 107 states and 164 transitions. It shows that we have just three possibilities to begin a sentence and it gives their tokens 170, 250 and 42 (these tokens are made-up digits for the paper).

This automaton is not fixed, it is changing (see §5). The transitions labeled by *NC* or *VC* refer to the two other automata, the noun phrase schema and the verb phrase schema. Figure 3 and Figure 4 present two schemata, very simplified, but enough for our example (in reality, the noun phrase schema has 205 states and 289 transitions and the verb phrase schema has 34 states and 70 transitions).

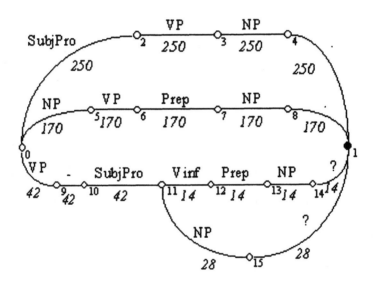

Fig. 2. An example of sentence schema

3   The Data Base

3.1   The Dictionary

Our dictionary is based on two dictionaries:

- The Juilland dictionary [4] that is made up of about 18 000 inflected words (5 083 lemmas), covering 92.43% of words in a text;
- The Catach dictionary [1] that roughly is a subset of the previous one and is made up of only 4 000 inflected words (1 620 lemmas), yet covering 90.51% of words in a text.

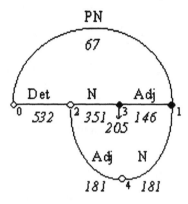

Fig. 3. An example of noun phrase schema

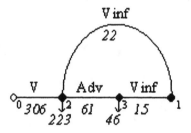

Fig. 4. An example of verb phrase schema

Of course, the 10% of remaining errors represent about two non predicted words by sentence, which is a huge quantities. So, this basis dictionary will be increased with new words, adapted to the user (§5).

The main role of the dictionary is to manage the token of the user's current words. We store two kinds of tokens: the token of lemma and the token of inflections. So, we correct the low appearance of some inflections of current words. First, the real token used is the one of lemma; second, we draw a distinction between the different inflections.

With the same example, assume that we are looking for the word *aime* (*like*). Table 1 presents the tokens of all its inflections that are in our dictionary and the token of the lemma (the sum of the other one).

Inflections		
aime	Present, 1s	110
aime	Present, 3s	57
aimer	Infinitive	42
aimait	Imperfect, 3s	31
...		
aimeront	Future, 3p	1
Lemma		
aimer		398

Table 1. Extract of dictionary

3.2   The Acceptability Tables

Definition 2 *One names acceptability table, a binary matrix M defined by the succession of two terms:*

- *if the succession of the term of the row i and the term of the column j is acceptable:*
  $M[i, j] = +$
- *else:*
  $M[i, j] = -$

[6] [7].

The *HandiAS* system includes three acceptability tables, one for every occurrence schema. They tell us the possible successions between the different syntactical categories.

Still with the same example, Table 2 presents an acceptability table for categories recognized by the noun phrase schema of Figure 3. One can read on this table:

- a proper noun can begin or ending a noun phrase
- a determiner can begin but not ending a noun phrase and can be followed by a noun or an adjective

4   The Working of the System

Now, we explain how the system works with the beginning of same sentence: *Tu aimes beaucoup la musique. (You like music a lot.).*

	Begining of phrase	Proper noun	Det	Noun	Adj	End of phrase
Proper Noun	+	−	−	−	−	+
Determiner	+	−	−	+	+	−
Noun	−	−	−	+	+	+
Adjective	−	−	−	+	+	+

Table 2. An acceptability table

4.1 Before the First Letter

Before writing, we are at state 0 of the sentence schema with three tokens:

$$\lambda(0, NC) = 170, \ \lambda(0, SubjPro) = 250, \ \lambda(0, VC) = 42$$
$$\text{and, of course:}$$
$$\Sigma_{l \in L} \lambda(0, l) = 462$$

and, at state 0 of the noun phrase schema and the verb phrase schema:

$$\lambda(0, PN) = 67, \ \lambda(0, Det) = 532, \ \Sigma_{l \in L} \lambda(0, l) = 599$$
$$\lambda(0, V) = 306, \ \Sigma_{l \in L} \lambda(0, l) = 306$$

and, thus, we compute four probabilities of syntactical categories:

$$P_{state0}(NC \wedge Det) = 170/462 * 532/599 \approx 0.33$$
$$P_{state0}(NC \wedge PN) = 170/462 * 67/599 \approx 0.04$$
$$P_{state0}(SubjPro) = 250/462 \approx 0.54$$
$$P_{state0}(VC \wedge V) = 42/462 \approx 0.09$$

After, we consult the dictionary and we compute the probabilities of lemmas (Table 3).

And we put forward a list of five words: *le, la, les, l'* (lemma *le, the*), *il* (lemma *il, he*).

4.2 After the First Letter

When the user writes the first letter, *t*, we compute only the words that begin by this letter. So, we put forward a new list of five words: *tu, t'* (lemma *tu, you*), *ton, ta, tes* (lemma *ton, your*).

As soon as the user chooses the word *tu*, we write it in the text screen, with a space after. Thus, he has done two actions (to click on *t* and *tu*) instead of three actions (to click on *t*, *u* and *space*). It is not so bad, because this word has very few letters. We will see on §6 that *HandiAS* writes a word after two or three letters.

Syntactical category		Lemma		$P_{state0}(Phrase \wedge Cat \wedge Lemma)$
PN	55	Jean	17	$17/55 * P_{state0}(NC \wedge PN) \approx 0.01$
		...	...	...
Det	54074	le	38585	$38585/54074 * P_{state0}(NC \wedge Det) \approx 0.24$
		un	13839	$13839/54074 * P_{state0}(NC \wedge Det) \approx 0.08$
		...	...	...
SubjPro	24481	il	10851	$10851/24481 * P_{state0}(SubjPro) \approx 0.24$
		je	8774	$8774/24481 * P_{state0}(SubjPro) \approx 0.19$
		tu	4856	$4856/24481 * P_{state0}(SubjPro) \approx 0.11$
		...	...	...
V	63547	être	13190	$13190/63547 * P_{state0}(NC \wedge V) \approx 0.02$
		...	...	...

Table 3. Lemma probabilities

5 Evolution and Adaptation

The most important points of the HandiAS system are its evolution and adaptation to user's vocabulary and syntax. Without this evolution and this adaptation, the data of HandiAS (dictionary and token schemata) are too poor to be efficient. But, we tailor a software to the individual. And the regular vocabulary and syntax of the user are going to make complete the data. Of course, the keyboarding of a known word (in the dictionary or in the new word list) leads to automatically increment its token. The tokens on token schematon do the same.

Now, we are explaining how we deal with new words, for instance *rap*, in the sentence *Jean aime le rap.* (*John loves the rap.*). This word is not in the dictionary. We are on state 2 of the noun phrase schema (Figure 3) and we read two transitions:

$$\lambda(2, N) = 351, \lambda(2, Adj) = 181, \Sigma_{l \in L}\lambda(2, l) = 532$$

With two probabilities:

$$P_{state2}(N) = 351/532 \approx 0.66$$
$$P_{state2}(Adj) = 181/532 \approx 0.34$$

So, we suppose that *rap* is probably a noun, may be an adjective... And nothing else, as we see on acceptability table (Table 2). The occurrence of the word *rap* later evolves like other words of dictionary. However, between the current session, this token is increased, so that a rare word, before used in a text, may reappear in suggestion lists.

At the end of current session, *HandiAS* will propose its hypothesis on the word *rap* to the user. After validation, the word will be put on dictionary. If the user intends to never use it also, he can choose to refuse this addition.

The syntax evolves too. Assume that one wants to write *Où as-tu appris le rap?* (*Where do you learn rap?*). This sentence begin by a coordinating conjunction, then it continues by an auxiliary verb (*avoir*), a subject pronoun, a past participle and a noun phrase, before ending by a question mark. After checking that this phrase structure is acceptable in an acceptability table, the sentence schema (Figure 2) will be modify as on Figure 5.

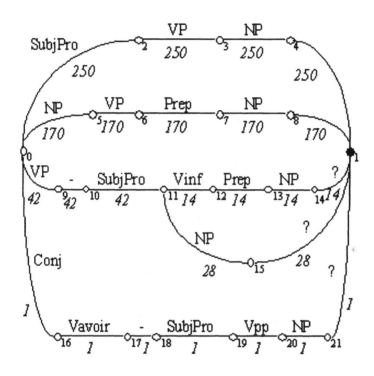

Fig. 5. Modification of sentence schema

6 Results and Prospects

Today, the *HandiAS* system is going to implement in C++ language and Windows NT system. A software company [2] gives with the Handimousse package a free version of the interface of HandiAS without suggestions, just virtual keyboard and annouces the market of the real HandiAS system for next January.

---

[2] C Technologie, 2 rue Marie-Curie, F44470 Carquefou, France

However, we have realized tests with a corpus before the beginning of this new implementation. Table 4 gives the results of these tests [9]. The total number of these actions must be compared with the 10 866 characters of the corpusthat we have used for the tests. With these three tests, we note that the HandiAS system (dictionary and token schemata) gives best result than a by itself token dictionary system or a by itself token schemata system.

		Only with token dictionary	Only with schema system	HandiAS	Without it
	Selections	3 680	3 653	3 345	0
Actions	Keyboarding	3 290	2 953	2 791	10 866
	TOTAL	6 970	6 606	6 136	10 866
% of doing actions		64.15	60.80	56.47	100
% of saved actions		35.85	39.20	43.53	0

Table 4. Test results

An other important test is to know the number of letters keyboarding before writting a word. Table 5 shows that our system is yet more efficient than a by itself token dictionary or a by itself schema system [5].

With our system		
Only with token dictionary	Only with schema system	HandiAS
3.13	2.90	2.61

Table 5. Number of keyboarding letters to write a word

After the realization of this prototype, we think adapt our system to other language, because the implementation of *HandiAS* is independent of the language. We need only to build an other linguistic data base.

References

1. Catach N. (1984), *Les listes orthographiques de base du français*, Paris, Nathan.
2. Debili F. (1982), *Analyse syntaxico-sémantique fondée sur une acquisition automatique de relations lexicales sémantiques*, Professoral thesis, University Paris 11.
3. Harris Z. S. (1968), *Mathematical Structures of Language*, Interscience Publishers

4. Juilland A., Brodin D., Davidovitch C. (1970), *Frequency dictionary of french words*, La Haye, Mouton & Co.

5. Le Pvdic B. (1997), *Prédiction Morphosyntaxique évolutive dans un système d'aide à la saisie de textes pour des personnes handicapées physiques*, Doctoral thesis, University of Nantes.

6. Maurel D. (1991), Préanalyse des adverbes de date du français, *TA information*, volume 32-2, p. 5-17.

7. Maurel D. (1996), Building automaton on Schemata and Acceptability Tables, *First Workshop on Implementing automata (WIA 96)*, London, Ontario, in *Lecture Notes in Computer Science*, vol. 1260, 72-86.

8. Maurel D., Chauvier L. (1997), Pseudo-minimum transducers : Building Minimum Subsequential Finite-state Transducers, *Second Workshop on Implementing automata (WIA 97)*, London, Ontario, in *Lecture Notes in Computer Science*, vol. 1436, 122-132.

9. Maurel D., Le Pévédic B., Yavchitz J. (1998), La prédiction lexicale et syntaxique à partir du contexte gauche : Application au système *HandiAS*, *Revue Informatique et Statistique dans les Sciences Humaines*, vol. 33.

10. Mohri M. (1993), *Analyse et représentation par automates de structures syntaxiques composés*, Doctoral thesis, University Paris VII.

11. Mohri M. (1997), Finite-state transducers in language and speech processing, *Computational Linguistics*, vol. 23-2, 269-311.

12. Roche E. (1993), *Analyse syntaxique transformationnelle de français par transducteurs et lexique-grammaires*, Doctoral thesis, University Paris VII.

Bi-directional Automata to Extract Complex Phrases from Texts

Thierry Poibeau

THOMSON-CSF/LCR
Domaine de Corbeville
91404 Orsay Cedex, France
poibeau@lcr.thomson-csf.com

Abstract. This paper presents an experiment to develop natural-language tools to improve the quality of documents. These softwares are using finite-state automata enriched with notions of proximity, optionality and contextual information. They are called bi-directional because they need to parse a sequence not only from the left to right-hand side of a sentence, but on both sides of a word. This method improves efficiency.

Keywords:
Automata, natural language processing, design and architecture of automata software, industrial applications, digital documents

1 Introduction: Analyzing Corporate Documents

[1]Every document issued by a company is intended to produce a reaction. For this reason, the author and those associated with its production are implicated in the document sent out. In doing so the company (depending on its activity) is running risks which range from time wasted to important failures. It can also mean financial loss and major dysfunctioning, for example, in the sending out of costing and technical specifications[2].

---

[1] This work has been done while the author was working in NEMESIA (Ivry-sur-Seine). Information about NEMESIA can be found at the following address: http://www.nemesia.com. NTK.FOCUS is a trademark from NEMESIA and can be downloaded and tested in different versions at the above address. I would like to thank people from NEMESIA for their constant friendship. I want to thank Adeline Nazarenko (LIPN), Grégory Perrat (NEMESIA), Marie-Paule Péry-Woodley (ERSS) and two anonymous reviewers from WIA'98 for their comments and suggestions on earlier versions of this paper.

[2] Most of this work has been done in close collaboration with people working for the nuclear industry. In this domain, the risk factor must be strictly controlled, especially the documentation.

To be efficient, textual documents must then satisfy quality criteria and respect pre-determined stylistic norms. Moreover, they should use standardized and systematic vocabulary, refer to the dictionary of existing terms, avoid the use of fuzzy or imprecise terms, and the overuse of synonyms [4]. But, if we can find spelling and grammar checkers, stylistic and content checkers are lacking.

Our aim is then to develop small tools integrated in the working environment of experts. Our point of view is that, as far as is possible, the verification of quality criteria must be adapted to the rereading of documents, thus facilitating the proof-reading and the annotation of the documents analyzed.

We present here a versatile formalism to represent complex expressions in texts, including morphological variations, word reordering and surface transformations. This formalism is intended to be integrated in tools to control various criteria concerned with fuzzy expressions, requirement analysis and terminological norms. For different reasons that will be discussed, we chose to implement enriched automata, that is to say automata able to reflect notions like proximity, optionality, etc. They permit an efficient analysis of texts and increase the quality of documents.

2   The Risk Factor in Corporate Documents

The introduction of new document technology (e.g. groupware, workflow, Intranet, ...) requires a preliminary analysis of risks inherent in the document [12] [13].

2.1   Controlling the Risk Factor

This task is achieved by:

- identifying documents sent out or processed by the corporate and the accompanying risks,
- identifying the authors and the readers of the documents,
- analyzing the distribution plan for the document,
- identifying the criteria to be applied to different types of documents depending on their inherent risks.

These criteria, defined in the context of major risk, are flexible enough to be adapted to all possible risks.
We are particularly concerned with documentation of projects that becomes overwhelming while teams spend more and more time writing it up. This is the only information which survives the projects. However, this documentation is never written with a view to future [17]. Defining quality control procedures for project documents is achieved by identifying the following:

- the types of documents depending on the phases of the project and their future uses,

- the writers, their availability at each stage and their responsibility within the project,
- quality criteria to be applied to different types of documents.

A project, whatever it may be, lives and dies according to its documentation. To guarantee quality is to remove one of the main sources of incoherence and misunderstanding harmful to good communication between those working on the project. The means of validating technical documents, which accompany a technical project, principally concern those documents which are standardized or partially standardized. When it comes to validating non-standardized language, which is its weak point, analytical techniques become subjective and intuitive, which is the direct opposite to the demands of quality.

We developed a method called AVIS to analyze technical texts and identify their inherent risks [12]. This method was partially implemented in the AVISO toolbox (a big lisp program running under UNIX). We are now extracting some parts of this toolbox and re-implementing them through small tools devoted to specific tasks and integrated in the standard working environment of documentation experts.

2.2 NTK.FOCUS: A Tool to Search Fuzzy Expressions in Texts

We developed a tool named NTK.FOCUS, to search fuzzy expressions in technical texts [14]. A fuzzy expression is an expression that creates doubt about the content of a text. This doubt is revealed by written terms which leave too much open to interpretation or which seem to be incomplete or imprecise, and this may well lead to an incorrect understanding of the text. Information which is too vague amounts to non-information.

A detailed linguistic study of this phenomenon has enabled us to determine different types of fuzziness. In order to detect fuzziness, five types of linguistic markers have been identified and these point out[3]:

- a missing element (e.g. in an enumeration): *etc.*
- an element which leads to uncertainty and gives no further information: *peut-être*[4].
- an element which leads to uncertainty while giving more information: *probablement*[5].
- an element which is fuzzy in a given context: *plus* (the fuzziness in *il faut plus de sodium*[6] but not in *il ne faut plus de sodium*[7]).
- a fuzzy element in relation to a given norm or field: *normal* (when the normal/abnormal characteristics of the event in question have not been specified).

---

[3] This work is fully reported in [14].
[4] Tr.: *maybe.*
[5] Tr.: *probably.*
[6] Tr.: *more sodium is needed.*
[7] Tr.: *no more sodium is needed.*

The development of our dictionaries is cyclic. At the beginning, a collaboration of experts and linguists permits to establish a first set of fuzzy expressions. A first dictionary is then available and validated on real texts. An analysis of the results is made in terms of noise (expressions that should not be recognized) and silence (expressions that should be recognized). Dictionaries are then corrected and validated on new corpus, etc. This work is defined on a linguistic basis and our methodology permits to avoid problems related to complete introspective systems. Resulting texts must avoid ambiguity, respect industrial standards and be easily translated in formal systems[8]. NTK.FOCUS functions in the same way as other tools integrated in word processors[9]. It is interactive because it can be used while working on the document : it enables the user to make corrections and the re-reader to insert footnotes. It is ultimately up to the writer (or the re-reader) to judge whether the expression in the text is acceptable or not. He can correct, add notes or ignore it.

This tool has been developed in close collaboration with re-reader of technical documents, in particular experts from the energy industry. Every document concerning nuclear energy must undergo several verification stages. The elimination of fuzzy expression is one of the first stage in the verification cycle. NTK.FOCUS is currently used, among all, by experts from the French Institute of Protection and Nuclear Safety (IPSN, Institut de Protection et de Sécurité Nucléaire, Fontenay-aux-Roses, France) and from the French energy supplier EDF (Electricité de France, Clamart).

3   Implementing Bi-directional Automata

Our aim in this context is to build a versatile formalism to represent complex expressions in texts including morphological variations, word reordering and surface transformations. This formalism is intended to be integrated in tools to control various criteria concerned with fuzzy expressions, requirement analysis and terminological norms. For reasons of efficiency and disposability, we exclude any pre-processing or tagging of the text. Automata are a good formalism for our purpose [15] [19]: although automata do not offer a deep analysis of texts, they are very efficient and robust.

3.1   Morphological Variations of Words

We separate each word into a stem, a prefix and a suffix. A suffix family corresponds to the variation set of a stem. We could have taken in consideration

---

[8] Experiments have been done at IPSN to translate natural language descriptions of complex systems in K.O.D. (Knowledge Oriented Design) [21] or in the formal B language [2]. This means that the texts must not be too much open to interpretation nor be incomplete or imprecise.

[9] NTK.FOCUS is fully integrated in Microsoft Word 6, 7 and 8 (Microsoft Word 97). Dictionaries for French, English and German have been developed.

some etymological or phonological criterion to break down words [9]. For example, let's consider the French stem *possible* (Figure 1) which can only be suffixed by $(\varepsilon + s + ment)$[10].

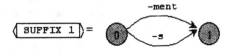

Fig. 1. Morphological variation of the stem *possible. The entrance point is state 0, exit points are grayed (states 0 or 1)*

Now, if we take a look at the stem *certain* (Figure 2), we can see that it can be suffixed by $(\varepsilon + s + e + es + ement)$. We could say that the suffix family of possible is the same as the one of *certain*. The variation is only due to the fact that possible has a vowel stem while certain has a consonant one: the -e- is then analyzed as an epenthetic vowel appearing before consonant suffixes:

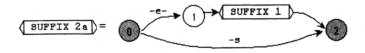

Fig. 2. Morphological variation of the stem *certain* (1)

But our aim is more pragmatic: it is easier and more efficient to represent a suffix family by an automata composed of a list of endings instead of a composition of automata (Figure 3). This is the automata representing the suffix family of *certain*. This automata is strictly equivalent of the one of figure 2.

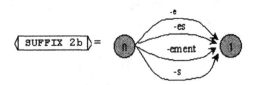

Fig. 3. Morphological variation of the stem *certain* (2)

On figure 3, one can notice that state 0 is the entry point of the automata but is also a final state. The stem can stay without any suffix. Finally, we add

---

[10] ε represents an empty string.

a tag on the suffix family to say if it is optional or not. The automata is then equivalent to a simple list of suffixes (Figure 4)

certain:	*lex='certain'*
	suffix=suffix1
suffix1:	*optional=yes*
	num=4
	lex=('e','s','es','ement')

Fig. 4. Morphological variation of the stem *certain* (3)

The same is done with prefixes. However, for reasons of efficiency, we generally prefer to duplicate the entry in order to have only empty prefixes[11]. Languages such as German set another problem because they have verbs with particles that can be detached in the sentence. For instance, *ausgehen (to go out) : ich gehe mit meinem Vater aus (I am going out with my father), ich will mit meinem Vater ausgehen (I want to go out with my father), ich bin mit meinem Vater ausgegangen (I have been out with my father)*. When the stem is very eroded like in the last example, it is often necessary to create a second entry for the word. This problem is solved by the way we represent multi-word entries.

Although this is a simplified implementation of morphological variations, dictionaries for French, German and English have been developed. Our aim was to build an efficient and realist way to implement morphology, not to describe the very complexity of morphology. For a more complete implementation and some considerations on the problem, see [7] [18].

3.2   Word Combinations in Complex Expressions

Classically, automata describe complex expressions in a single direction, from the left to the right-hand side of a sentence. This is known to be very efficient if you are looking for full words, because there is no backtracking during the analysis. If you are looking for markers (such as complex determiners or prepositional phrases), a lot of wrong analyses will be initiated, because most of the time markers begin with a preposition or a determiner, i.e with words among the most frequent in European languages.

To avoid this problem, we have defined in each automata a word called pivot. The pivot is a word from which every item of the expression is described . This

---

[11] So that a stem always begins on the left-hand side of a word. Some optimizations can then be applied such as a hash table.

means we can search for a full word instead of an empty one such as a preposition. We have defined rules to choose the pivot: it must be a full word and it is generally the semantic head of the expression.

Other items appearing in the description are described in terms of relative position (a word is on the left or on the right of anotherone), of proximity (there can be at most 1 or 2 words between the item described) and of optionality (a multi-word expression can be recognized with or without certain words). The description language allows to say that a word M is on the left of a word M', that they can be separated by at most n words, and that M is or not optional in the expression. Finally, two operators have been added to the description language: the OR operator (disjunction, if a position can be indifferently filled by more than one lexical item) and the AND operator (conjunction, if several lexical items are required on both sides of a word or if their ordering is free).

Let's take an example with the family of expressions built around the French word *peu*, in analogy with the notion of tree family in Tree Adjoining Grammars [1]. We find on figure 5 prepositional phrases and complex determiners expressing an idea of quantification.

Pivot	Expression
Peu	*peu*
Peu	*quelque peu*
Peu	*de peu*
Peu	*peu à peu*
Peu	*à peu près ...*

Fig. 5. Family of expressions built around peu (extract)

The description uses a bracketed formalism defined according to the aforementioned constraints. A dictionary is a set of descriptions, each one being equivalent to a bi-directional finite state automaton, enriched with information on proximity and optionality (Figure 6). Although bi-directional automaton are not so much used, it is a well-known formalism already described in [5]. It makes it possible to look for words at the same time on the left and the right-hand side of a word. Our description language is relatively rich and can offer several strategies to encode a same expression. However, some rules have been defined to keep descriptions coherent and explicit. The dictionary is then compiled for reasons of efficiency [3].

Herein, the dictionary is equivalent to a local grammar [7] [20]. However the concept of local grammar is not sufficient to describe some of the expressions that must be recognized according to the experts (see above, part 2).

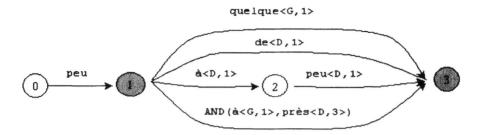

Fig. 6. A part of the bi-directional automaton corresponding to the pivot peu. *Apart from the pivot peu, each transition refers to a lexical item with indications on its position (compared to the item from the preceding transition) and a number referring to the proximity. The AND(à <1,G>, près <D,3>) expression allows to recognize "à peu près", where the pivot "peu" is between "à" and "près". Proximity on the transition from "peu" to "près" is equal to 3 to be able to recognize phrases such as "à peu de choses près" with insertions between "peu" and "près". This automaton is able to recognize all the expressions enumerated above.*

4 Context Sensitive Automata

Several tools have been developed with this formalism to increase corporate document quality. But, while we were developing NTK.FOCUS, it appeared that the formalism had to be enriched with context and inhibiting information.

The formalism was thus enriched in order to be able to take the context into account and particularly the frames in which the expression loses its fuzzy character (See the example above: *Il faut plus de sodium* vs. *Il ne faut plus de sodium*). The description of the expression is still in conformity with what has been expressed *supra*. But words are encapsulated in another structuration level depending on the context. A positive boolean feature is associated with the word if it expresses fuzziness, and a negative one when it reveals an inhibiting context. This automaton is in fact a transducer generating a Boolean value. Our representation reflects the way we implemented it: the Boolean result of the parsing is calculated and percolated during the analysis. However, this is not equivalent to a weighted transducer. A weighted transducer permits to increase the efficiency of parsing by the addition of heavy weights on the most common paths. Our aim here is not to increase the efficiency of parsing, but to be able to say that, in a certain context, an expression is no more fuzzy and must not be displayed in the interface (Figure 7).

To recognize a fuzzy expression, the analyzer must be on an automaton final state with the positive feature being active. If the negative feature is active, the analysis fails. See [10] [16] for problems caused by multiword expressions. The only output produced by an automata is generally a Boolean value saying if the

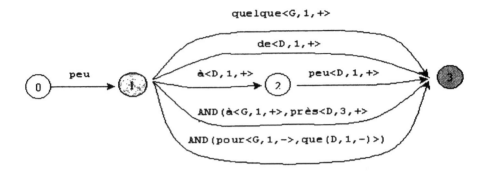

Fig. 7. Enriched automata with notion of context. *This is the same automaton as on Figure 6, where each transition corresponding to a fuzzy expression is tagged with a +. On the opposite, the transition corresponding to "pour peu que" ("Pour peu que'" is a locution meaning "if") is tagged with -, because pour peu que does not constitute a fuzzy expression ("pour peu qu'il soit malade..." "Tr.: If he is sick, ..."). Only the most extended sequence is recognized by the algorithm to avoid recognizing "peu" in the expression "pour peu que".*

segment of text has been recognized (parsed) or not. Contextual information associated with Boolean values permits to limit the number of segment of text appearing in the interface (to limit the noise) and to increase the quality of the results. This have been proved through experiments with experts from IPSN who compared a first version of NTK.FOCUS and the present version. The accuracy of the result is good according to them and leads to a better quality of the expert analysis over the document. A stack is used to parse automata and manage backtracking.

The analysis functions by success/failure: the analyzer looks first for fuzzy expressions, then for their negative contexts. When an expression is found with the positive feature being active, it is displayed in the interface. The analysis fails if the automata has been completely parsed and if no expression have been found with the positive feature being active.

5   Further Developments

We are now working on more complex redundant structures. For example requirement analysis need to parse not only markers, but also modal verbs and qualification adjectives. Modal verbs can occur with a lot of expressions and cannot be manually integrated in each automata. We have then developed of a pre-processing stage to expand our dictionary before compilation. This preprocessor is a script using the C precompiler and the programming language Sed. At the moment, this tool is being used to develop new dictionaries concerning requirement analysis [11]. Such overlapping automata can lead to very large entries (up to 300 lines and more for certain complex entries) that cannot be processed manually.

During our experiments on corpora, it appeared that the notion of sentence is not so clear as it seems [6]. That is not without consequences, particularly for requirement analysis. We added to the system a flexible module of text segmentation to be able to take lists and enumerations into account. It is then possible to extract meronymic relations from texts. for example, objects in technical texts are often described in terms of decomposition. One main task of the experts is to see if the text respects a model of the activity described. This means building semi-automatically models and knowledge bases to project texts on these models. Our current experiments are being tested in this context at IPSN.

6 Conclusion

We have shown in this paper finite-state automata enriched with notions of proximity, optionality and contextual information. These automata are called bi- directional because they need to parse a sequence not only from the left to right-hand side of a sentence, but on both sides of a word. This method improves efficiency: the first element searched is most of the time a full word instead of a preposition or a determiner.

We have also increased analysis precision by adding information about contextual inhibiting items. It is then possible to search an expression in a certain context but not in another one. Negative context is a good way to reduce noise in analysis results. Contextual information is completely integrated in the lexical description to ensure coherence.

Lastly, we have tried to show that automata can be used for general purposes, particularly for tools concerned with natural language processing. They offer a well-known, efficient and versatile formalism that is a real alternative to all ad hoc formalisms developed for particular purposes.

References

1. Abeillé A.: Les nouvelles syntaxes. Ed. A. Colin (Paris), (1993)
2. Abrial J.-R.: The B-book: assigning programs to meanings. Cambridge University Press (Cambridge) (1996)
3. Aho A., Sethi R., and Ullman J.: Compilers Principles, Techniques, and Tools. Ed. Addison-Wesley, (1986)
4. ANSSC (1967).: Code of good practices for the documentation of digital computer programs. American Nuclear Society Standards Committee, (1967)
5. Chomsky N. and Miller G.: Introduction to the formal analysis of natural languages. in Luce R.D., Bush R. & Galanter E., Handbook of Mathematical Psychology Wiley (New York), (1963) 269-322
6. Grefenstette G. and Tapanainen P.: What is a word, what is a sentence ? Problems of tokenization. COMPLEX (Budapest), (1994)

7. Gross M.: Méthodes en syntaxe. Ed. Hermann (Paris), (1975)

8. Gross M. and Perrin D. Electronic dictionaries and automata in computational linguistics. Lecture Notes in Computer Science **377**, Springer Verlag (Berlin) (1989)

9. Huot H.: Sur la notion de racine. TAL, vol bf 35 num **2**, (1994) 46-76

10. Karttunen L., Chanod J.-P., Grefenstette G. and SchillerA.: Regular expressions for language engineering. Natural Language Engineering, vol. **2** num **4**, (1996) 305-328

11. Kozlowska-Heuchin R.: L'analyse de la notion d'exigence dans un document technique en vue d'une extraction de connaissance automatisée. Bulag (actes de FRAC-TAL'97, Besançon), (1997) 227-234

12. Lippold B., Pomian J., Henry J.Y. and Elsensohn O.: AVIS : une méthode d'analyse de la cohérence des documents. Proceedings of the European Safety and Reliability conference (La Baule), (1994) 888-899

13. Lippold B. and Pomian J.: AVIS : une méthode d'analyse de la cohérence des documents techniques. Proceedings of the Fifteenth International Conference IA 95 - Language Engineering 95 (Montpellier), (1995) 301-312

14. Lippold B., Kozlowska-Heuchin R. and Poibeau T.: NTK.FOCUS : un outil d'aide à la rédaction des documents techniques. JST-FRANCIL'97 (Avignon), (1997) 369-374

15. Maurel D.: Reconnaissance de séquences de mots par automates, adverbes de date du français. Thèse de doctorat en informatique, Université de Paris 7 (1989)

16. Poibeau T. and Maurel D.: A la fin de : preposition ou determinant complexe dans les adverbiaux de temps ? Cahiers de grammaire, num **20**, (1995) 101-111

17. Pomian J.: Mémoire d'entreprise. Sapientia (Ivry), (1996).

18. Revuz D.: Dictionnaires et lexiques Méthodes et algorithmes. Thèse de doctorat en informatique, Université de Paris 7, (1991)

19. Roche E. and Schabes Y.: Finite State Language Processing. MIT Press(Cambridge), (1997)

20. Silberztein M.: Dictionnaires électroniques et analyse automatique des textes. Ed. Masson (Paris), (1993)

21. Vogel C.: Génie cognitif. Ed. Masson (Paris), (1988)

A Fast New Semi-incremental Algorithm for the Construction of Minimal Acyclic DFAs

Bruce W. Watson

[1] Department of Computer Science
University of Pretoria
Pretoria 0002, South Africa
watson@RibbitSoft.com
www.cs.up.ac.za
[2] Ribbit Software Systems Inc
(IST Technologies Research Group)
Box 24040, 297 Bernard Ave. Kelowna, B.C., V1Y 9P9, Canada
watson@RibbitSoft.com
www.RibbitSoft.com

Abstract. We present a semi-incremental algorithm for constructing minimal acyclic deterministic finite automata. Such automata are useful for storing sets of words for spell-checking, among other applications. The algorithm is semi-incremental because it maintains the automaton in near-minimal condition and requires a final minimization step after the last word has been added (during construction).

The algorithm derivation proceeds formally (with correctness arguments) from two separate algorithms, one for minimization and one for adding words to acyclic automata. The algorithms are derived in such a way as to be combinable, yielding a semi-incremental one. In practice, the algorithm is both easy to implement and displays good running time performance.

1 Introduction

In this paper, we present a semi-incremental algorithm for constructing minimal acyclic deterministic finite automata (ADFAs). By their acyclic nature, they represent finite languages, and are therefore useful in applications such as storing words for spell-checking. In such applications, the automata can grow extremely large (with more than 10^6 states), and are difficult to store without first applying a minimization procedure. In traditional minimization techniques, the unminimized ADFA is first constructed and then minimized. Unfortunately, the unminimized ADFA can be very large indeed — sometimes even too large to fit within the virtual address space of the host machine. As a result, incremental techniques for minimization (ie. the ADFA is minimized during its construction) become interesting. Incremental algorithms frequently have /* some /* overhead — if the unminimized ADFA fits easily within physical memory, it is still faster to use nonincremental techniques. On the other hand, with very large ADFAs, the incremental techniques may be the /* only /* option.

J.-M. Champarnaud, D. Maurel, D. Ziadi (Eds.): WIA'98, LNCS 1660, pp. 121–132, 1999.

The algorithm presented in this paper is /* semi /*-incremental (as opposed to /* fully /*-incremental, or just /* incremental /*) because it maintains the ADFA in a nearly-minimal condition while words are added, but requires a simple 'final' step to achieve full minimality after all words have been added.

In order to derive the algorithm, we proceed in three stages:

1. Derive an efficient algorithm for minimizing ADFAs.
2. Derive a simple algorithm for adding new words to an ADFA under certain conditions.
3. Combine the algorithms derived in the first two steps. By-design, the first two algorithms will manipulate the ADFA in a fashion that makes them combinable.

A great deal of practical work on minimizing ADFAs has been done. Unfortunately, much of the research is of a proprietary nature and thus forms part of the folklore of computational linguistics. Several years ago, Revuz also derived incremental minimization algorithms [3]. The primary algorithm presented by Revuz uses a reverse ordering of the words to quickly compress the endings of the words within the dictionary. Further work by Revuz has yielded algorithms which correspond rather closely to the one in this paper. All minimization algorithms show strong similarities, as can be seen from the taxonomy in [4]. The subtle differences between the algorithms can lead to domain-specific performance advantages for each algorithm. More recently, a paper describing "An Incremental Algorithm for Constructing Acyclic Deterministic Transducers" was accepted to the /* Workshop on Finite State Machines in Natural Language Processing /* [1]; the current author is also one of the co-authors of that paper. The algorithm presented in that paper is entirely different from the one presented in this paper. In particular, the following differences are noteworthy:

- The algorithm presented here is semi-incremental, whereas the two algorithms presented in the other paper are fully-incremental.
- The algorithm presented here requires the words (to be added to the ADFA) to be in /* any /* order of decreasing length. One of the algorithms appearing in the other paper requires the words in lexicographical order while the other algorithm does not have an ordering requirement.
- Thanks to the simple ordering requirement of the algorithm presented in this paper, the algorithm is both very simple (and thus easy to implement) and very fast.
- Also, thanks to the simplicity, full correctness arguments are more readily provided for the algorithm in this paper.

This paper is structured as follows:

- §2 presents the necessary definitions of ADFAs.
- §3 derives a procedure for minimizing an ADFA.
- §4 derives a procedure for adding words to an ADFA.
- §5 combines the algorithms from §3 and 4 to yield the semi-incremental algorithm.

- §6 provides some details on running time and implementation issues for the algorithm.
- §7 presents the conclusions of this paper.

2 Preliminaries

In this paper, we consider acyclic deterministic finite automata (ADFAs). The algorithm is readily extended to work with acyclic deterministic $/*$ transducers $/*$, though such an extension is not considered here.

A deterministic finite automaton (DFA) is a 6-tuple $(Q, \Gamma, \delta, q_0, F)$ where:

- Q is the finite set of states.
- Γ is the input alphabet. We choose this instead of the more traditional Σ since we will use that letter for 'summation' in the section on running time analysis.
- $\delta \in Q \times \Gamma \longrightarrow Q \cup \{\perp\}$ is the transition function. It is a partial function, and we use $\perp$ to designate the invalid state.
- $q_0 \in Q$ is the start state.
- $F \subseteq Q$ is the set of $/*$ final $/*$ states.

To make some definitions simpler, we will use the shorthand Γ_q to refer to the set of all alphabet symbols which appear as out-transition labels from state q. Formally,

$$\Gamma_q = \{\, a \mid a \in \Gamma \wedge \delta(q, a) \neq \perp \,\}$$

All of the algorithms presented in this paper are in the form of the guarded command language, a type of pseudo-code — see [2].

3 Minimizing an ADFA

In this section, we derive a procedure for minimizing an ADFA. We begin with an abstract algorithm (whose correctness is easily determined) and refine the abstract details to yield an efficient algorithm.

The primary definition of minimality of an ADFA M (indeed, this definition applies to *any* DFA, not just acyclic ones) is:

(for all M', such that M' is equivalent to M, $|M| \leq |M'|$ holds)

where equivalence of DFAs means that they accept the same language. This definition is difficult to manipulate (in deriving an algorithm), and so we consider one written in terms of the right languages of states. The right language of a state q, written $\overrightarrow{\mathcal{L}}(q)$, is the set of all words spelled out on paths from q to a final state. Using right languages (and the Myhill-Nerode theorem), minimality can also be written as the following predicate (which we call postcondition R):

(for all $p, q \in Q$, such that $p \neq q$, $\neg\boldsymbol{Equiv}(p, q)$ holds)

where we define predicate *Equiv* to be equivalence of states:

$$Equiv(p, q) \equiv \overrightarrow{\mathcal{L}}(p) = \overrightarrow{\mathcal{L}}(q)$$

(Additionally, we require that there are no useless states, though this additional restriction is not usually written and we ignore it in the rest of this paper since our algorithms have no way of creating useless states.)

To achieve postcondition R, we introduce a procedure *minimize* which:

- assumes the ADFA as a global data-structure;
- takes (as first parameter) a set of states U, called the *unique* states, which are pairwise inequivalent; that is:

  (for all $p, q \in U$, such that $p \neq q$, $\neg Equiv(p, q)$ holds)

- does not shrink the set U of pairwise inequivalent states;
- takes (as second parameter) another set of states V (which is disjoint from U, ie. $U \cap V = \emptyset$) which are to be made pairwise unique - those which are not unique will be removed since they are redundant.

To be concise, we define part of our invariant, I_1, as the predicate:

(for all $p, q \in U$, such that $p \neq q$, $\neg Equiv(p, q)$ holds)

In the following presentation of the algorithm, we do not give all of the shadow variables or the complete (and lengthy) invariants for a full correctness argument:

---

proc *minimize*(U, V)
{invariant: $U \cap V = \emptyset \wedge I_1$}
{variant: —V—}
while $V \neq \emptyset$ **do**
  let $q : q \in V$
  $V := V - \{q\}$
  if there exists p, **such that** $p \in U$ $Equiv(p, q)$ **then**
    {q is redundant}
    let $p : p \in U \wedge Equiv(p, q)$
    redirect all of q's in-transitions to p
    remove state q, since it's redundant
  else
    {q is unique}
    $U := U + \{q\}$
  end if
end while
{$V = \emptyset$}

Note that the invariant is also a precondition of the entire procedure. Invoking the procedure as *minimize*$(\emptyset, Q)$ would clearly minimize the entire ADFA.

The main difficulty with this algorithm is testing *Equiv*. This algorithm would actually be applicable to DFAs with cycles, if we found a way of practically implementing the guard; see [4, Chapter 7] for a wide variety of algorithms (including an incremental one) that are usable on DFAs with cycles. Fortunately, we can use an inductive definition of $\overrightarrow{\mathcal{L}}$:

$$\overrightarrow{\mathcal{L}}(q) = \left[\bigcup_{a \in \Gamma_q} \{a\} \overrightarrow{\mathcal{L}}(\delta(q,a)) \right] \cup \begin{cases} \{\varepsilon\} & \text{if } q \in F \\ \emptyset & \text{if } q \notin F \end{cases}$$

Intuitively, a word z is in $\overrightarrow{\mathcal{L}}(q)$ if and only if

- z is of the form az' where $a \in \Gamma$ is a label of an out-transition from q to $\delta(q,a)$ (ie. $a \in \Gamma_q$) and z' is in the right language of $\delta(q,a)$, or
- $z = \varepsilon$ and q is a final state.

We can begin rewriting *Equiv* as follows (where each line of the rewriting is separated by a /* hint /* corresponding to the rewriting step):

$Equiv(p, q)$

$\equiv$ { definition of *Equiv* }

     $\overrightarrow{\mathcal{L}}(p) = \overrightarrow{\mathcal{L}}(q)$

$\equiv$ { the inductive definition of $\overrightarrow{\mathcal{L}}$ }

     $(\varepsilon \in \overrightarrow{\mathcal{L}}(p) \equiv \varepsilon \in \overrightarrow{\mathcal{L}}(q)) \wedge \Gamma_p = \Gamma_q \wedge$

     (for all a, such that $a \in \Gamma_p \cap \Gamma_q$, $\{a\} \overrightarrow{\mathcal{L}}(\delta(p,a)) = \{a\} \overrightarrow{\mathcal{L}}(\delta(q,a))$ holds)

$\equiv$ { the inductive definition of $\varepsilon \in \overrightarrow{\mathcal{L}}(p)$ }

     $(p \in F \equiv q \in F) \wedge \Gamma_p = \Gamma_q \wedge$

     (for all a, such that $a \in \Gamma_p \cap \Gamma_q$, $\{a\} \overrightarrow{\mathcal{L}}(\delta(p,a)) = \{a\} \overrightarrow{\mathcal{L}}(\delta(q,a))$ holds)

$\equiv$ { for two languages L_0, L_1: $(\{a\}L_0 = \{a\}L_1) \equiv (L_0 = L_1)$ }

     $(p \in F \equiv q \in F) \wedge \Gamma_p = \Gamma_q \wedge$

     (for all a, such that $a \in \Gamma_p \cap \Gamma_q$, $\overrightarrow{\mathcal{L}}(\delta(p,a)) = \overrightarrow{\mathcal{L}}(\delta(q,a))$ holds)

$\equiv$ { definition of *Equiv* }

     $(p \in F \equiv q \in F) \wedge \Gamma_p = \Gamma_q \wedge$

     (for all a, such that $a \in \Gamma_p \cap \Gamma_q$, $Equiv(\delta(p,a), \delta(q,a))$ holds)

Clearly, the last step has yielded a recursive definition which, while implementable (since we have /* acyclic /* DFAs and the recursion will end), is not very efficient (interestingly, this would lead to the algorithm in [4, Section 7.4.6]). Fortunately, the acyclicity also yields a way to implement this efficiently by restricting the invariant in the algorithm (as in the following paragraphs).

The evaluation of $Equiv(p, q)$ would be much simpler if we were assured that all $\delta(p, a)$ and $\delta(q, a)$ were already in our unique set U (ie. the children $\delta(p, a)$ and $\delta(q, a)$ are pairwise unique). Since we will use this requirement extensively, we write it as predicate $P_1(r)$ (for state r):

(for all a, such that $a \in \Gamma_r$, $\delta(r, a) \in U$ holds)

Ideally, we could assume $P_1(p) \wedge P_1(q)$, in which case we could rewrite

$$P_1(p) \wedge P_1(q) \wedge Equiv(\delta(p,a), \delta(q,a))$$
$\equiv$ { definition of $Equiv$ }
$$P_1(p) \wedge P_1(q) \wedge \overrightarrow{\mathcal{L}}(\delta(p,a)) = \overrightarrow{\mathcal{L}}(\delta(q,a))$$
$\equiv$ { property of U stated in invariant I_1 and assumption $P_1(p) \wedge P_1(q)$ }
$$P_1(p) \wedge P_1(q) \wedge \delta(p,a) = \delta(q,a)$$

Here, by introducing the assumption of $P_1(p) \wedge P_1(q)$, we have successfully removed the recursion in *Equiv*. Of course, it remains to ensure that this assumption holds. We consider the two assumed conjuncts separately.

To ensure that $P_1(q)$, we introduce an invariant I_2 (which is also happens to be a precondition) which states that:

(for all r, a, such that $r \in V \wedge a \in \Gamma_r$, $\delta(r,a) \in V$ or $\delta(r,a) \in U$ holds)

(Note that $U \cup V$ is not necessarily equal to Q, so this is not as simple as it looks.) A predicate such as I_2 is trivially true if r has no out-transitions or there is no $r \in V$ (ie. $V = \emptyset$). Given I_2 and the acyclic property of ADFAs, we know that for any $r \in V$, there will be a chain of r's children eventually ending with one in U. From this, we conclude that we can always choose /* some /* $q \in V$ such that $P_1(q)$.

Turning to $P_1(p)$, we introduced another invariant I_3:

(for all r, such that $r \in U$, $P_1(r)$ holds)

We can now summarize our derivation from the last two derivations above:

$$P_1(p) \wedge P_1(q) \wedge (p \in F \equiv q \in F) \wedge \Gamma_p = \Gamma_q \wedge$$
(for all a, such that $a \in \Gamma_p \cap \Gamma_q$, $Equiv(\delta(p,a), \delta(q,a))$ holds)
$\equiv$ { invariants I_2 and I_3, assumption $P_1(p) \wedge P_1(q)$, previous derivation }
$$P_1(p) \wedge P_1(q) \wedge (p \in F \equiv q \in F) \wedge \Gamma_p = \Gamma_q \wedge$$
(for all a, such that $a \in \Gamma_p \cap \Gamma_q$, $\delta(p,a) = \delta(q,a)$ holds)

We label the latter three conjuncts (ie. excluding the assumption) of the last predicate $P_2(p,q)$. Given this, we can rewrite our procedure.

This algorithm is still invoked as $minimize(\emptyset, Q)$.

Because of I_2 and the $P_1(q)$ assumption, this algorithm operates in a bottom-up fashion (i.e. from final states towards the start state q_0, if we imagine q_0 at the top of the page, with the final states towards the bottom) on the set V, where (by precondition I_2) U is 'hanging' below V.

Instead of V being a set, we can reinforce this 'bottom-up' order by stipulating that V is a stack with those items on the top of the stack being topologically 'lower' (closer to U, and thus further from q_0) than the items lower in the stack — ie. if r is on top of the stack, then $P_1(r)$.

Before restructuring the algorithm to use a stack, we rewrite V as a stack in invariant I_2 (I_3 remains the same). I'_2, states that:

```
proc minimize(U, V)
{invariant: U ∩ V = ∅ ∧ I₁ ∧ I₂ ∧ I₃}
{variant: | V | }
while V ≠ ∅ do
   let q : q ∈ V ∧ P₁(q)
   V := V − {q}
   if there exists p, such that p ∈ U, P₂(p, q) then
      {q is redundant}
      let p : p ∈ U∧ P₂(p, q)
      redirect all of q's in-transitions to p
      remove state q, since it's redundant
   else
      {q is unique}
      U := U + {q}
   end if
end while
{V = ∅}
```

**(for all r, a, such that $r \in V \land a \in \Gamma_r$, ($\delta(r, a)$ is higher in V,
or $\delta(r, a) \in U$) holds)**

This gives the algorithm (where [] is the empty stack):

To minimize the entire ADFA, we invoke it with $minimize(\emptyset, V)$, where V is a preorder stacking of the states Q (/* Any /* preorder will do, since the out-transitions from any given state are unordered).

4   Adding Words to an ADFA

At first glance, in adding a word w to an ADFA we would simply trace out the path (corresponding to w), from the start state q_0, and make the resulting state a final one (add it to F). Unfortunately, this may inadvertently add more than one word to the ADFA in the event that some state on the path has more than one in-transition.

Under certain conditions, we can, however, use such a simple procedure. A sufficient condition is that, along the w path there must be no state q with more than one in-transition. We can state this condition formally as $P_3(w)$:

(for all i, such that $0 \leq i < |w|$, $pred(\delta^*(q_0, w_{[0..i)})) \leq 1$ holds)

where $pred(q)$ is the number of /* predecessors /* (in-transitions) of state q and δ^* is the usual reflexive and transitive closure of the transition function δ.

Condition P_3 is trivially guaranteed if we are adding word to a /* trie /* — a tree-structured ADFA. We state condition P_3 because we will be simultaneously minimizing the ADFA, and so we may have a situation where not all states will have at most one predecessor.

```
proc minimize(U, V)
{invariant: U ∩ V = ∅ ∧ I₁ ∧ I′₂ ∧ I₃}
{variant: | V | }
while V ≠ [] do
   q := pop(V)
   {P₁(q)}
   if there exists p, such that p ∈ U, P₂(p, q) then
      {q is redundant}
      let p : p ∈ U ∧ P₂(p, q)
      {Equiv(p, q)}
      redirect all of q's in-transitions to p
      remove state q, since it's redundant
   else
      {q is unique}
      U := U + {q}
   end if
end while
{V = []}
```

We can now give the procedure *add_word* which takes the word, adds it to the ADFA (which is taken to be a global data-structure) and returns the state at which the word ends (**cand** is conditional conjunction):

(*build_state* is a function for extending the ADFA with a new transition. It returns the newly created state.) The first loop proceeds as far as possible in the existing ADFA, while the second loop extends the ADFA, as necessary, to accommodate the rest of w.

5   Combining the Algorithms

If we are simultaneously minimizing and adding words, we must co-ordinate the two algorithms. Recall, from the last version of the minimizing algorithm that only states in U have been minimized (and may, therefore, have more than one predecessor) and that set U is grown bottom-up (towards q_0). To synchronize the algorithms, it makes sense to add the word-set W such that their final states are also added bottom-up.

One possible way to order W is to add the words in any order of decreasing Length. We say /* any /*, since there are usually several such possible orderings. In this case, while adding w, we will not pass through any final states, and it will be safe to minimize the portion of the ADFA below the top-most (closest to q_0) final states. We call these top-most final states the /* upper final states frontier /* since they form the set of final states which appear first over all paths leading away from q_0. The states below this frontier will become our minimization set U.

After invoking *add_word(w)*, the states at and below the returned state (inclusive) are safe for minimization and can be added to the stack (in preorder)

```
func add_word(w) : Q
{P₃(w)}
q, i := q₀, 0
{invariant: q = δ*(q₀, w[0..i))}
{variant: | w | −i }
while i <| w | cand δ(q, wᵢ) ≠⊥ do
   q, i := δ(q, wᵢ), i + 1
end while
if i <| w | then
   {δ(q, wᵢ) =⊥}
   while i <| w | do
      q, i := build_state(q, wᵢ), i + 1
   end while
else
   skip
end if
F := F ∪ {q}
return q
```

using this procedure (which initially takes the returned value of add_word and the empty stack):

```
proc build_stack(q, X) : Stack of states
push(X, q)
for a : a ∈ Γ_q do
   if δ(q, a) ∈ F then
      skip
   else
      X := build_stack(δ(q, a), X)
   end if
end for
return X
```

Along each q-to-leaf path, this procedure stops when it encounters another final state (which, by our invariant, will already be in U).

After adding word w, we minimize the following stack:

$$build\_stack(add\_word(w), [\,])$$

After adding all of the words, we still need to minimize the states above the upper final states frontier. This is done with an additional invocation of add_word and minimize.

This yields our final, combined, semi-incremental algorithm:

{we have an empty ADFA with a single start state}
$U := \emptyset$
{invariant: U is the upper final states frontier and below}
{variant: $|\,W\,|$ }
while $W \neq \emptyset$ **do**
    let $w : w \in W \wedge w$ is *any* shortest word in W
    $W := W - \{w\}$
    $minimize(U,\ build\_stack(add\_word(w),[]))$
end while
{all have been minimized except those above the final states frontier}
$minimize(U, build\_stack(q_0,[]))$
{R}

6   Implementation and Performance

We begin with an analysis of the running time of the final algorithm. We have the following sub-analyses:

- The outer repetition executes exactly $|W|$ times. We will actually ignore this and separately calculate the total running time of the function invocations.
- For any word w, $add\_word(w)$ adds $\mathcal{O}(|w|)$ states and takes the same order time.
- Each state is pushed onto the stack exactly once (thereafter it is placed in U or removed due to redundancy, and by acyclicity never appears in the stack again).
- An invocation $minimize(U, X)$ takes $\mathcal{O}(|\Gamma| \cdot |X|)$ time. The factor $|\Gamma|$ is due to the test $P_2(p, q)$, while—X—is due to the outer loop of that procedure.

Here, we have assumed that some elementary operations (such as set membership in U, stack operations and automata transitions) can be done in constant time. The total time taken by add_word (and also the total number of states) is

$$\mathcal{O}(\sum_{w:w \in W} |w|)$$

This also happens to be the time taken (in total) by $build\_stack$ and the order of the total stack size.

Given that each state is pushed onto the stack exactly once, the total time taken by $minimize$ is

$$\mathcal{O}(|\Gamma| \cdot \sum_{w:w \in W} |w|)$$

It follows that the total running time of this algorithm is:

$$\mathcal{O}(|\Gamma| \cdot \sum_{w:w \in W} |w| + \sum_{w:w \in W} |w| + \sum_{w:w \in W} |w|)$$

or, equivalently

$$\mathcal{O}(|\Gamma| \cdot \sum_{w:w \in W} |w|)$$

Interestingly, this running time is asymptotically the same as the running time of the best known non-incremental algorithms for ADFAs (they are also linear in the size of the ADFA, whose construction is in-turn linear in $\sum_{w:w \in W} |w|$). Naturally, the semi-incrementality takes its toll in the constant factor.

In practical terms, for performance reasons, we can do the following:

- U can be made global.
- U can be indexed by a state's finality.

Complete performance data for this algorithm is not yet available. Preliminary benchmarking shows, however, that it is substantially faster (sometimes by a factor of 2.3) than Ribbit Software Systems Inc.'s implementations of the two algorithms presented in [1]. Those algorithms maintain full minimality (thereby incurring the additional cost). Benchmarking data used to-date (with the algorithm in this paper) have shown that the ADFA remains within 27% of minimal (in terms of state count).

7   Conclusions

We have presented a simple semi-incremental algorithm for acyclic DFAs — indeed, it appears to be the first such algorithm published. The following aspects of the algorithm are noteworthy:

- The algorithm is particularly simple and easily understood.
- The simplicity of the algorithm goes hand-in-hand with the formal derivation and presentation of correctness arguments.
- In order to formally derive a (semi-)incremental algorithm, novel techniques were used, such as: separately developing component algorithms that are nondeterministic and using their nondeterminism to synchronize them (cooperatively) into the incremental one.
- The running time of the algorithm is asymptotically as good as the best non-incremental ones.

Acknowledgements: Nanette Saes and Richard Watson were kind enough to proofread this paper. The referees also provided a great number of useful comments, and Jan Daciuk's recent work inspired me to write up this algorithm.

References

1. Daciuk, J.D., Watson, B.W. and R.E. Watson. An Incremental Algorithm for Constructing Acyclic Deterministic Transducers. (Proceedings of the International Workshop on Finite State Methods in Natural Language Processing, Ankara, Turkey, 30 June–1 July 1998).

2. Dijkstra, E.W. *A Discipline of Programming.* (Prentice Hall, Englewood Cliffs, N.J., 1976).

3. Revuz, D. *Dictionnaires et lexiques: méthodes et algorithmes.* (Ph.D dissertation, Institut Blaise Pascal, LITP 91.44, Paris, France, 1991).

4. Watson, B.W. *Taxonomies and Toolkits of Regular Language Algorithms.* (Ph.D dissertation, Eindhoven University of Technology, The Netherlands, 1995). See `www.RibbitSoft.com/research/watson`

Using Acceptors as Transducers

Matti Nykänen*

Department of Computer Science
University of Helsinki
P.O. Box 26 (Teollisuuskatu 23) FIN-00014
matti.nykanen@cs.helsinki.fi

Abstract. We wish to use a given nondeterministic two-way multi-tape acceptor as a transducer by supplying the contents for only some of its input tapes, and asking it to generate the missing contents for the other tapes. We provide here an algorithm for determining beforehand whether this transduction always results in a finite set of answers or not. We also develop an algorithm for evaluating these finite answers whenever the previous algorithm indicated their existence. Our algorithms can also be used for speeding up the simulation of these acceptors even when not used as transducers.

1 Introduction

In this paper we study the following problem: assume that we are given a nondeterministic two-way multi-tape acceptor $\mathcal{A}$ and a subset X of its tapes. We would like to use $\mathcal{A}$ no longer as an acceptor, which receives input on all its tapes, but as a kind of transducer [7, Chapter 2.7] instead, which receives input on tapes X only, and generates as output the set of missing inputs onto the other tapes. We then face the following two problems:

1. Can it be guaranteed that given any choice of input strings for tapes X the set of corresponding outputs of $\mathcal{A}$ will always remain finite?
2. Even if problem 1 could be solved positively, how can the actual set of outputs corresponding to a given choice of input strings be computed?

Our motivation for studying these problems arises from *string databases* [3,5,9], which manipulate strings instead of indivisible atomic entities. Such databases are of interest for example in bioinformatics, because they allow the direct representation and manipulation of the stored nucleotide (DNA or RNA) sequences.

If we assume an SQL-like notation [1, Chapter 7.1] for the query language, then one possible query for such a string database might be stated as follows.
SELECT x_2
    FROM x_1 IN R
    WHERE $\phi_{\mathrm{rev}}(x_1, x_2)$

---

* Supported by the Academy of Finland, grant number 61382.

Here $\phi_{\text{rev}}(x_1, x_2)$ is some user-defined expression which compares the strings w_1 and w_2 denoted by the variables x_1 and x_2, say "w_2 is the reversal of w_1". Then this query requests every string w_2 that is the reversal of some string w_1 currently stored in the database table R. Note in particular that these strings w_2 need (and in general can) not be stored anywhere in the database; the query evaluation machinery must generate them instead as needed.

We have developed elsewhere [5,10] a logical framework for such a query language. This framework accommodates expressions like $\phi_{\text{rev}}(x_1, x_2)$ via a multidimensional extension of the modal Extended Temporal Logic suggested by Wolper [17]. The multi-tape acceptors studied here are exactly the computational counterparts to these logical expressions.

A given query to a database is considered to be *safe* for execution if there is a way to evaluate its answer finitely. One safe plan for evaluating the aforementioned query would be as follows, where $L(\mathcal{A}_{\text{rev}})$ is the string relation accepted by $\mathcal{A}_{\text{rev}}$, a multi-tape acceptor corresponding to the expression $\phi_{\text{rev}}(x_1, x_2)$. (One such acceptor is shown as Fig. 1 below.)

> **for all** strings w_1 in table R **do**
> $\quad V \leftarrow \{w_2\colon \langle w_1, w_2 \rangle \in L(\mathcal{A}_{\text{rev}})\};$
> $\quad$ **output** every string in V
> **end for**

Our two problems stem from these safe evaluation plans. Problem 1 is "How could we infer from $\phi_{\text{rev}}(x_1, x_2)$ that the set V is always going to be finite for every string w_1 that can appear in R?" Problem 2 is in turn "Even if this is so, how can we simulate this $\mathcal{A}_{\text{rev}}$ (efficiently) for each w_1 to generate these V?" (We have studied elsewhere [4][10, Chapter 4.4] how solutions to problem 1 guide the selection of safe execution plans.)

One possible solution would have been to restrict the language for the expressions $\phi_{\text{rev}}(x_1, x_2)$ beforehand into one, which ensures this finiteness by definition, say by fixing x_1 to be the input variable, which is mapped into the output variable x_2 as a kind of transduction [3,9]. However, in logic-based data models [1], the use of transducers seems less natural than acceptors, because the concept of information flow from input to output is alien to the logical level, and of interest only in the query evaluation level. But we must eventually also evaluate our string database queries, and then we must infer which of our acceptors can be used as transducers, and how to perform these inferred transductions, and thus we face the aforementioned problems.

The rest of this paper is organized as follows. Sect. 1.1 presents the acceptors we wish to use as transducers. Sect. 1.2 formalizes problem 1, and reviews what is known about its decidability. Sect. 2 presents our algorithm, which gives a sufficient condition for answering problem 1 in the affirmative. Sect. 3 presents then an explicit evaluation method for those acceptors that this algorithm accepts, answering problem 2 in turn. Finally, Sect. 4 concludes our presentation.

1.1   The Automaton Model

Let the alphabet Σ be a finite set of characters. We also assume *left and right tape end-markers* '[' *and* ']' not in Σ. Then we define the nth character of a given string $w \in \Sigma^*$ with length $|w| = m$ as

$$\underbrace{c_1 \ldots c_m}_{w}[n] = \begin{cases} [& n = 0 \\ ] & n = m+1 \\ c_n & 1 \leq n \leq m. \end{cases}$$

Intuitively our automaton model is a "two-way multi-tape nondeterministic finite state automaton with end-markers"; similar devices have been studied by for example Harrison and Ibarra [6] and Rajlich [11]. Formally, a *k-tape Finite State Automaton (k-FSA)* [5, Section 3][10, Chapter 3.1] is a tuple $\mathcal{A} = \langle Q_\mathcal{A}, s_\mathcal{A}, F_\mathcal{A}, T_\mathcal{A} \rangle$ with four elements, where (I) $Q_\mathcal{A}$ is a finite set of *states*; (II) $s_\mathcal{A} \in Q_\mathcal{A}$ is a distinguished *start* state; (III) $F_\mathcal{A} \subseteq Q_\mathcal{A}$ is the set of *final* states; and (IV) $T_\mathcal{A}$ is a set of *transitions* of the form $p \xrightarrow[d_1,\ldots,d_k]{c_1,\ldots,c_k} q$, where $p, q \in Q_\mathcal{A}$, each $c_i \in \Sigma \cup \{[,]\}$, and each $d_i \in \{-1, 0, +1\}$. We moreover require that $d_i = -1$ implies $c_i \neq [$ and $d_i = +1$ implies $c_i \neq]$; this ensures that the heads do indeed stay within the tape area limited by these end-markers.

A *configuration* of $\mathcal{A}$ on *input* $w = \langle w_1, \ldots, w_k \rangle \in (\Sigma^*)^k$ is of the form $C = \langle p, n_1, \ldots, n_k \rangle$, where $p \in Q_\mathcal{A}$ and $0 \leq n_i \leq |w_i| + 1$ for all $1 \leq i \leq k$. This C corresponds intuitively to the situation, where $\mathcal{A}$ is in state p, and each head $i = 1, \ldots, k$ is scanning square number n_i of the tape containing string w_i. Hence we say that $\langle q, n_1 + d_1, \ldots, n_k + d_k \rangle$ is a possible *next* configuration of C if and only if $p \xrightarrow[d_1,\ldots,d_k]{w_1[n_1],\ldots,w_k[n_k]} q \in T_\mathcal{A}$. Now $+1$ can be interpreted as "read forward", while -1 means "rewind the tape to read the preceding square again", and 0 "stand still". We call tape i of $\mathcal{A}$ *unidirectional* if no transition in $T_\mathcal{A}$ specifies direction -1 for it; otherwise tape i is called *bidirectional* instead.

A *computation* of $\mathcal{A}$ on input w is a sequence $\mathcal{C} = C_0 C_1 C_2 \ldots$ of these configurations, which starts with the *initial* configuration $C_0 = \langle s_\mathcal{A}, 0, \ldots, 0 \rangle$, and each C_{j+1} is a possible next configuration of the preceding configuration C_j. This computation $\mathcal{C}$ is *accepting* if and only if it is finite, its last configuration C_f has no possible next configurations, and the state of this C_f belongs to $F_\mathcal{A}$. The *language* $L(\mathcal{A})$ *accepted by* $\mathcal{A}$ consists of those inputs w, for which there exists an accepting computation $\mathcal{C}$.

Because $\mathcal{A}$ is nondeterministic, we can without loss of generality assume that no transitions leave the final states $F_\mathcal{A}$. We can for example introduce a new state $f_\mathcal{A}$ into $Q_\mathcal{A}$, and set $F_\mathcal{A} = \{f_\mathcal{A}\}$. Then for every state p previously in $F_\mathcal{A}$, and every character combination $c_1, \ldots, c_k \in \Sigma \cup \{[,]\}$, on which there is no transition leaving p, we add the transition $p \xrightarrow[0,\ldots,0]{c_1,\ldots,c_k} f_\mathcal{A}$. In this way, whenever a computation of $\mathcal{A}$ would halt in state p, it performs instead one extra transition into the (now unique) new final state $f_\mathcal{A}$, and halts there.

We often view $\mathcal{A}$ as a *transition graph* $G_\mathcal{A}$ with nodes $Q_\mathcal{A}$ and edges $T_\mathcal{A}$. In particular, a computation of $\mathcal{A}$ can be considered to trace a path $\mathcal{P}$ within $G_\mathcal{A}$

starting from node s_A. It is furthermore expedient to restrict attention to *non-redundant* $\mathcal{A}$, where each state is either s_A itself or on some path $\mathcal{P}$ from it into some state in F_A. Fig. 1 presents a 2-FSA $\mathcal{A}_{\text{rev}}$ in this form, where $\Sigma = \{a, b\}$. $L(\mathcal{A}_{\text{rev}})$ consists of the pairs $\langle u, v \rangle$, where string v is the reversal of string u: looping in state II finds the right end of the bidirectional tape 1 without moving the unidirectional tape 2, while looping in state III compares the contents of these two tapes in opposite directions.

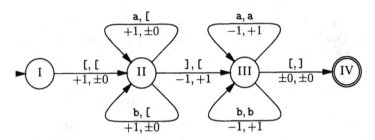

Fig. 1. A 2-FSA for Recognizing Strings and Their Reversals.

Another often useful simplification is the following.

Definition 1 *Tape i of k-FSA $\mathcal{A}$ is locally consistent if and only if every consecutive pair*

$$p \xrightarrow[d_1,\ldots,d_k]{c_1,\ldots,c_k} q \xrightarrow[d'_1,\ldots,d'_k]{c'_1,\ldots,c'_k} r$$

of transitions in T_A satisfies the condition

$$(d_i = 0 \Rightarrow c'_i = c_i) \wedge (d_i = +1 \Rightarrow c'_i \neq [) \wedge (d_i = -1 \Rightarrow c'_i \neq]). \tag{1}$$

This ensures that there are configurations, in which this pair can indeed be taken; whether these configurations do ever occur in any computation is quite another matter. For example, both tapes in Fig. 1 are locally consistent. If in particular tape i is both unidirectional and locally consistent, then given any path

$$\mathcal{P} = s_A \xrightarrow[d_{(1,1)},\ldots,d_{(k,i)}]{c_{(1,1)},\ldots,c_{(k,1)}} p_1 \xrightarrow[d_{(1,2)},\ldots,d_{(k,2)}]{c_{(1,2)},\ldots,c_{(k,2)}} p_2 \xrightarrow[d_{(1,3)},\ldots,d_{(k,3)}]{c_{(1,3)},\ldots,c_{(k,3)}} p_3 \cdots p_m$$

in G_A we can construct an input string $w_i = c'_1 c'_2 c'_3 \ldots c'_m$ for tape i, which allows $\mathcal{P}$ to be followed, if we choose

$$c'_j = \begin{cases} c_{(i,j)} & \text{if } (d_{(i,j)} = +1 \vee j = m) \wedge (c_{(i,j)} \neq [), \text{ and} \\ \varepsilon & \text{otherwise.} \end{cases} \tag{2}$$

For example in Fig. 1 w_2 spells out the sequence of transitions taken when looping in state III. Harrison and Ibarra provide a related construction for deleting unidirectional input tapes from multi-tape pushdown automata [6, Theorem 2.2], while Rajlich [11, Definition 1.1] allows the reading head to scan two adjacent squares at the same time for similar purposes.

Again the nondeterminism of $\mathcal{A}$ allows us to enforce Definition 1, at the cost of expanding the size of $\mathcal{A}$ by a factor of $(|\Sigma| + 2)$: construct a k-FSA $\mathcal{B}$ with state space $Q_{\mathcal{B}} = Q_{\mathcal{A}} \times (\Sigma \cup \{[,]\})$, which remembers the character under tape head i. Add for each transition $p \xrightarrow[d_1,\dots,d_k]{c_1,\dots,c_k} q$ the transitions $\langle p, c_i \rangle \xrightarrow[d_1,\dots,d_k]{c_1,\dots,c_k} \langle q, c_i' \rangle$ into $T_{\mathcal{B}}$ satisfying Eq. (1). Set finally $s_{\mathcal{B}} = \langle s_{\mathcal{A}}, [\rangle$ and $F_{\mathcal{B}} = F_{\mathcal{A}} \times (\Sigma \cup \{[,]\})$ to complete our construction.

1.2   The Limitation Problem

This section introduces our *limitation problem* [5, Section 4][10, Definition 3.3] concerning the automata defined in Sect. 1.1.

Definition 2 *Given a $(k + l)$-FSA $\mathcal{A}$, determine if there exists a limitation function $\mathcal{W} \colon \mathbb{N}^k \to \mathbb{N}$ with the following property: if $\langle u_1, \dots, u_k, v_1, \dots, v_l \rangle \in L(\mathcal{A})$, then $\max \{|v_j| : 1 \le j \le l\} \le \mathcal{W}(|u_1|, \dots, |u_k|)$.*

If this is the case, then we say that $\mathcal{A}$ satisfies the *finiteness dependency* [12] $\Gamma = \{1, \dots, k\} \rightsquigarrow \{k+1, \dots, k+l\}$. These dependencies are a special case of *functional* dependencies in database theory [1, Section 8.2]. Intuitively, $\mathcal{A}$ is a finite representation of the conceptually infinite database table $L(\mathcal{A})$, while Γ assures that if we select rows from this table by supplying values for the columns $1, \dots, k$, we do always receive a finite answer. In this way $\mathcal{A}$ can be safely used as a string processing tool within our string database model.

In terms of automata theory we require that for any *input* $\langle u_1, \dots, u_k \rangle \in (\Sigma^*)^k$ the possible *outputs* $\langle v_1, \dots, v_l \rangle \in (\Sigma^*)^l$ must remain a finite set. This is what is meant by "using acceptors as transducers": we supply strings for only some tapes (here $1, \dots, k$) of the acceptor $\mathcal{A}$, and ask it to produce us all those contents for the missing tapes (here $k + 1, \dots, k + l$) it would have accepted given the known tape contents. The limitation problem is then to determine beforehand whether this computation will always return a finite result or not. Weber [15,16] has studied a related question whether the set of *all* possible outputs on any inputs of a given transducer remains finite, and if so, what is the maximal output length.

The hardness of the limitation problem has been shown to depend crucially on the amount of bidirectional tapes in $\mathcal{A}$. The problem is undecidable for FSAs with *two* bidirectional tapes [10, Chapter 4.1]: given a Turing machine [7, Chapter 7] M one can write a corresponding 3-FSA $\mathcal{A}_M$ with two bidirectional tapes, which accepts exactly the tuples $\langle u, v, w \rangle$, where v and w together encode a sequence of computation steps taken by M on input u. Here v and w must be read twice, requiring bidirectionality. Then asking whether $\mathcal{A}_M$ satisfies $\{1\} \rightsquigarrow \{2, 3\}$ amounts to asking whether M is total.

The limitation problem becomes decidable if we restrict attention to those FSAs with at most *one* bidirectional tape [5, Theorem 4.2][10, Chapter 4.2]. Intuitively, all the unidirectional tapes are first made locally consistent, after which Eq. (2) allows us to construct their contents at will, so that we can concentrate on the sole bidirectional tape. This tape can in turn be studied by using an extension of the well-known *crossing sequence construction* [7, Chapter 2.6] for converting two-way finite automata into classical one-way finite automata. This method is clearly impractical, however. Therefore this paper presents in Sect. 2 a practical partial solution, which furthermore applies even in some cases involving multiple bidirectional tapes.

Example 1. The 2-FSA $\mathcal{A}_{\text{rev}}$ in Fig. 1 satisfies both $\{1\} \rightsquigarrow \{2\}$ and $\{2\} \rightsquigarrow \{1\}$ with the same limitation function $\mathcal{W}(n) = n$, because the reversal of a string is no longer than the string itself. This is moreover decidable, because only tape 1 is bidirectional in $\mathcal{A}_{\text{rev}}$. To see how limitation inference proceeds consider Fig. 2, which exhibits the crossing behavior of $\mathcal{A}_{\text{rev}}$ when tape 1 contains the string ab. For example, determining $\{2\} \rightsquigarrow \{1\}$ involves checking that every character written onto the bidirectional output tape 1 is "paid for" by reading something from the unidirectional input tape 2 as well, although this payment may occur much later during the computation; here it occurs when tape 1 is reread in reverse. This can in turn be seen from the automaton $\mathcal{B}$ produced by the crossing sequence construction by noting that the loops of $\mathcal{B}$ around the repeating crossing sequence indicated in Fig. 2 consume tape 2 as well.

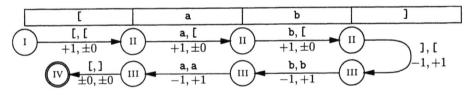

Fig. 2. A Crossing Behavior of the 2-FSA in Fig. 1.

The 2-FSA $\mathcal{A}_{\text{rev}}$ is also considered to satisfy the trivial finiteness dependency $\{1,2\} \rightsquigarrow \emptyset$ by definition. On the other hand, $\mathcal{A}_{\text{rev}}$ does not satisfy $\emptyset \rightsquigarrow \{1,2\}$, because $L(\mathcal{A}_{\text{rev}})$ is not finite.

2   An Algorithm for Determining Limitation

Our technique for solving the limitation problem given in Definition 2 is based on the following two observations. Let $\mathcal{A}$ be the $(k+l)$-FSA and $\Gamma = \{1, \ldots, k\} \rightsquigarrow \{k+1, \ldots, k+l\}$ the finiteness dependency in question.

1. If $\mathcal{A}$ accepts some input $\langle w_1, \ldots, w_{k+l}\rangle$ with some computation $\mathcal{C}$, where some head $k \leq j \leq k+l$ never visits the corresponding right end-marker ']', then $\mathcal{A}$ also accepts all the suffixed inputs

$$\langle w_1, \ldots, w_{j-1}, w_j \Sigma^*, w_{j+1}, \ldots, w_{k+l}\rangle$$

with the same $\mathcal{C}$. Hence, $\mathcal{A}$ cannot satisfy Γ in this case.

2. If on the other hand every accepting computation of $\mathcal{A}$ visits ']' on all output tapes, then the only way $\mathcal{A}$ can violate Γ is by *looping* while generating output onto some output tape but without consuming any of the inputs at the same time.

However, Sect. 1.2 indicated that reasoning about actual computations is infeasible. Thus we reason instead about the structure of the transition graph $G_{\mathcal{A}}$. Therefore, instead of observation 1, the algorithm in Fig. 3 merely tests that there is no path $\mathcal{P}$ from the start state $s_{\mathcal{A}}$ into a final state, which never requires ']' to appear on some output tape, whereas it would have sufficed to show that no such $\mathcal{P}$ is ever traversed during any accepting computation. ($\mathbb{B}$ denotes the Boolean type with values $\mathbf{0}$ as 'false' and $\mathbf{1}$ as 'true'.)

Similarly, the algorithm in Fig. 4 enforces a more stringent condition than observation 2: every cycle $\mathcal{L}$ in $G_{\mathcal{A}}$, during which some output tape is advanced to direction $+1$, must also move some input tape i into direction ± 1 but not back into the opposite direction ∓ 1. Then this tape i acts as a *clock*, which eventually terminates the potentially dangerous repetitive traversals of $\mathcal{L}$. Again, if $\mathcal{A}$ violates Γ, then some $\mathcal{L}'$ failing this condition must exist in $G_{\mathcal{A}}$, but the converse need not hold, because repetitions of $\mathcal{L}'$ need not necessarily occur during any accepting computation of $\mathcal{A}$. This condition is enforced by repeatedly deleting transitions, which cannot take part in any $\mathcal{L}'$ of this kind. This technique is related to analyzing the input-output behavior of logic programs [8,14], which analyze the call graph of the given program component by component.

More precisely, the edge deletions made by the algorithm in Fig. 4 can be justified as follows. Consider the first call made by the main algorithm in Fig. 5. Every loop mentioned in observation 2 must clearly be contained in some component H_i of $G = G_{\mathcal{A}}$, the entire transition graph of the k-FSA $\mathcal{A}$ being tested.

1. A transition between two different strongly connected components cannot then surely belong any such loop. The deletions in step 3 are therefore warranted.
2. Any transition τ that winds the clock tape j selected for the current component H_i cannot belong to any such loop either, because τ cannot be traversed indefinitely often. These traversals will namely wind the input tape j onto either end-marker, because tape j is not wound into the opposite direction by any other transition τ' in H_i. The deletions in step 6 are therefore warranted as well.

This reasoning can then be applied in the subsequent recursive calls on the reduced components H_i as well, because we can then assume inductively that the loops broken during the earlier calls could not have been ones mentioned in observation 2.

Formalizing this reasoning shows that Fig. 5 is indeed correct, as follows.

Theorem 1. *Let $\mathcal{A}$ be a $(p+q+r)$-FSA with tapes $1,\ldots,p$ unidirectional, and let the algorithm in Fig. 5 return $\mathbf{1}$ on $\mathcal{A}$ and $\{1,\ldots,p+q\}$. Then $\mathcal{A}$ satisfies $\{1,\ldots,p+q\} \rightsquigarrow \{p+q+1,\ldots,p+q+r\}$ with the function*

function halting(G:transition graph G_A of a k-FSA A;
$\qquad\qquad\qquad$ X:subset of $\{1,\dots,k\}$):$\mathbb{B}$;

1: $b \leftarrow 1$;
2: **for all** $i \in \{1,\dots,k\} \setminus X$ **do**
3: $\quad H \leftarrow G$ without transitions that specify ']' for tape i;
4: $\quad b \leftarrow b \wedge (H$ contains no path from s_A into any state in $F_A)$
5: **end for**
6: **return** b
end halting;

Fig. 3. An Algorithm for Testing Observation 1.

function looping(G:subgraph of G_A for a k-FSA A;
$\qquad\qquad\qquad$ X:subset of $\{1,\dots,k\}$):$\mathbb{B}$;

1: $b \leftarrow 1$;
2: $H_1,\dots,H_m \leftarrow$ the maximal strongly connected components of G;
3: Delete from G all transitions between different components (1.);
4: **for all** $i \leftarrow 1,\dots,m$ **do**
5: $\quad$ **if** some tape $j \in X$ winds to direction ±1 in H_i but not in direction ∓1 **then**
6: $\quad\quad$ Delete from H_i all transitions that wind this tape j (2.);
7: $\quad\quad d \leftarrow$ looping(H_i, X)
8: $\quad$ **else**
9: $\quad\quad d \leftarrow$ no tape in $\{1,\dots,k\} \setminus X$ winds into direction $+1$ in H_i
10: $\quad$ **end if**
11: $\quad b \leftarrow b \wedge d$
12: **end for**
13: **return** b
end looping;

Fig. 4. An Algorithm for Testing Observation 2.

function limited(A:k-FSA;
$\qquad\qquad\qquad$ X:subset of $\{1,\dots,k\}$):$\mathbb{B}$;

1: **return** halting$(G_A, X) \wedge$ looping(G_A, X)
end limited;

Fig. 5. An Algorithm for Determining Limitation.

$$\mathcal{W}(m_1, \ldots, m_p, n_1, \ldots, n_q) = g_{\mathcal{A}}(m_1, \ldots, m_p, n_1, \ldots, n_q) - 1, \; where$$

$$g_{\mathcal{A}}(m_1, \ldots, m_p, n_1, \ldots, n_q) = |Q_{\mathcal{A}}| \left(\max(p, 1) + \sum_{i=1}^{p} m_i \right) \left(\prod_{j=1}^{q} (n_j + 2) \right).$$

Proof. Let us assume that $\mathcal{C}$ is an arbitrary computation of $\mathcal{A}$ on some input

$$z = \langle u_1, \ldots, u_p, v_1, \ldots, v_q, w_1, \ldots, w_r \rangle.$$

We begin by proving the following two *claims* about this $\mathcal{C}$, which correspond to observations 1 and 2:

1. If $\mathcal{C}$ is accepting, then for every tape $p + q + 1 \leq j \leq p + q + r$ $\mathcal{C}$ takes some transition, which requires ']' on tape j.

Let otherwise h be a tape, which violates this claim 1. $\mathcal{C}$ traces a path through G_A from s_A into some state in F_A. Then step 4 of the algorithm in Fig. 3 sets $b = 0$ when testing $i = h$, which violates our assumption that the algorithm in Fig. 5 returns **1**, thus proving this claim 1.

2. No head $p + q + h$ moves to direction $+1$ more than

$$l = \mathcal{W}(|u_1|, \ldots, |u_p|, |v_1|, \ldots, |v_q|) + 1$$

   times during $\mathcal{C}$.

Assume to the contrary that some h violates this claim 2. Post a *fence* between two adjacent configurations C_g and C_{g+1} in $\mathcal{C}$ whenever tape $p + q + h$ moves to direction $+1$. By our contrary assumption, at least $l + 1$ of these fences are posted. Consider on the other hand two configurations C_x and C_y of $\mathcal{C}$ to have the same *color* if and only if they share the same state and the same head positions for tapes $1, \ldots, p + q$. At most l of these colors are available, recalling the assumption that tapes $1, \ldots, p$ are unidirectional. Therefore $\mathcal{C}$ must contain two configurations C_x and C_y, which have the same color, but are separated by an intervening fence. Consider then the sequence of transitions, which transform C_x into C_y, as a path $\mathcal{L}$ within G_A. This $\mathcal{L}$ forms a closed cycle, and tape heads $1, \ldots, p + q$ are on the same squares both before and after traversing $\mathcal{L}$, because C_x and C_y shared a common color. Let us then see which of the steps 3 or 6 of the algorithm in Fig. 4 will first delete some transition that belongs to $\mathcal{L}$. It cannot be step 3, because all of $\mathcal{L}$ belongs initially to the same maximal strongly connected component. But it cannot be step 6 either, because if $\mathcal{L}$ ever moves a tape $j \in X$ into some direction ± 1, it must also move tape j into the opposite direction ∓ 1 as well, in order to return its head onto the same square both before and after $\mathcal{L}$. Hence $\mathcal{L}$ persists untouched to the very end of the recursion on step 9, and there the presence of the transition of $\mathcal{L}$ that crosses the fence between C_x and C_y yields $d = \mathbf{0}$, which subsequently violates our assumption that the algorithm in Fig. 5 returns **1**, thus proving this claim 2.

Claims 1 and 2 are combined into a proof of the theorem as follows. Assume that $z \in L(\mathcal{A})$; that is, $\mathcal{A}$ has some accepting computation $\mathcal{C}$ on input z. It suffices to show that $|w_h| \leq l - 1$ for every $1 \leq h \leq r$. Tape head $p + q + h$ must cross every border between two adjacent tape squares from left to right, because otherwise $\mathcal{C}$ would not meet claim 1. Claim 2 states in turn that $\mathcal{C}$ performs at most l crossings of this kind. This means that tape $p + q + h$ contains at most $l + 1$ squares, of which the first and the last are reserved for the end-markers, leaving at most $l - 1$ squares for the characters of the input string w_h.   □

Example 2. Consider the 2-FSA $\mathcal{A}_{\text{rev}}$ in Fig. 1. The algorithm in Fig. 5 can detect that it satisfies $\{1\} \rightsquigarrow \{2\}$ as follows. The algorithm in Fig. 3 returns **1**, because every path into the final state IV must contain III $\xrightarrow[0,0]{[,]}$ IV.

Evaluating the algorithm in Fig. 4 proceeds in turn as follows. First, all transitions from one state into another are deleted in step 3, leaving only the loops around states II and III. This is depicted in Fig. 6, where the components themselves are dotted, and the transitions between them (and thus deleted in step 3) are dashed. These loops are in turn deleted in step 6 when processing the corresponding components, and therefore this function eventually returns **1** as well. However, Theorem 1 provides a rather imprecise limitation function $\mathcal{W}(n) = 4n + 7$ compared to the one given in Example 1.

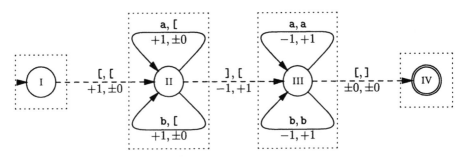

Fig. 6. The division of the 2-FSA in Fig. 1 into components.

On the other hand, the algorithm in Fig. 5 fails to detect $\{2\} \rightsquigarrow \{1\}$, which Example 1 did detect: looping in state II advances tape 1 without moving tape 2, and seems therefore dangerous to the algorithm in Fig. 4. Intuitively, $\mathcal{A}_{\text{rev}}$ first guesses nondeterministically some string, and only later verifies its guess against the input. Example 1 examined crossing sequences to see that this later checking in state III indeed reduces the acceptable outputs into only finitely many (here just one).

Essentially the same limitation function as in Theorem 1 suffices whenever all of the output tapes $p + q + 1, \ldots, p + q + r$ to be limited are unidirectional, even if $\{1, \ldots, p + q\} \rightsquigarrow \{p + q + 1, \ldots, p + q + r\}$ cannot be verified by the algorithm in Fig. 5 [4, Theorem 2.1]. This is natural, because the algorithm in Fig. 5 ignored the effects of moving any output tape $p + q + 1, \ldots, p + q + r$ into direction -1.

Theorem 2. *Let the only bidirectional tapes of the $(p + q + r)$-FSA $\mathcal{A}$ be $p + 1, \ldots, p + q$, and let $\mathcal{A}$ satisfy $\Gamma = \{1, \ldots, p + q\} \rightsquigarrow \{p + q + 1, \ldots, p + q + r\}$. Then*

$$W'(m_1, \ldots, m_p, n_1, \ldots, n_q) = (|\Sigma| + 2)^r g_{\mathcal{A}}(m_1, \ldots, m_p, n_1, \ldots, n_q) - 1$$

is a corresponding limitation function, where $g_{\mathcal{A}}$ is as in Theorem 1.

Proof. Consider the proof of Theorem 1, and assume further in claim 2 that all the output tapes $p + q + 1, \ldots, p + q + r$ are made locally consistent as in Definition 1; this new assumption introduces the factor $(|\Sigma| + 2)^r$ into W'. With this modification, the original fencing-coloring construction shows that if some accepting computation $\mathcal{C}$ on input z advances some output tape $p + q + h$ more than $W'(|u_1|, \ldots, |u_p|, |v_1|, \ldots, |v_q|) + 1$ times, then the path of transitions taken by this $\mathcal{C}$ can be partitioned into three sub-paths $\mathcal{KLM}$, where $\mathcal{L}$ begins in C_x and ends in C_y, which share the same color, and contains a transition τ that crosses some fence between C_x and C_y. However, now $\mathcal{A}$ must also accept all the pumped inputs

$$\left\langle u_1, \ldots, u_p, v_1, \ldots, v_q, w_{(1,\mathcal{K})} w_{(1,\mathcal{L})}^t w_{(1,\mathcal{M})}, \ldots, w_{(r,\mathcal{K})} w_{(r,\mathcal{L})}^t w_{(r,\mathcal{M})} \right\rangle,$$

where $t \in \mathbb{N}$, and each $w_{(k,\mathcal{J})}$ denotes the string of characters in those squares of output tape $p + q + k$, onto which the head lands during $\mathcal{J}$ (']' excluded). This is in effect an application of Eq. (1) to the output tapes $p + q + 1, \ldots, p + q + r$. The presence of τ within $\mathcal{L}$ shows that $w_{(h,\mathcal{L})} \neq \varepsilon$, and hence Γ fails by observation 2, thereby proving this modified claim 2.

Claim 1 continues to hold, as reasoned in observation 1, and the theorem again follows as before. □

Turning now to assessing when the algorithm in Fig. 5 does detect finiteness dependencies, we see that it is successful at least when all tapes are unidirectional.

Theorem 3. *Let $\mathcal{A}$ be a non-redundant $(k + l)$-FSA with all tapes unidirectional and locally consistent, and let the algorithm in Fig. 5 return $\mathbf{0}$ on $\mathcal{A}$ and $\{1, \ldots, k\}$. Then $\mathcal{A}$ does not satisfy $\Gamma = \{1, \ldots, k\} \rightsquigarrow \{k + 1, \ldots, k + l\}$.*

Proof. The non-redundancy of $\mathcal{A}$ and the unidirectionality and local consistency of all its tapes imply by Eq. (2) that for every path $\mathcal{P}$ in $G_{\mathcal{A}}$ we can always find an accepting computation $\mathcal{C}$ on some input w traversing $\mathcal{P}$. Letting $\mathcal{P}$ then be any subgraph of $G_{\mathcal{A}}$, which caused the algorithm in Fig. 5 to return $\mathbf{0}$ yields some $\mathcal{C}$, whose existence violates Γ along the lines of Theorem 2. □

3 Evaluation of the Limited Answers

Having inferred the $(k+l)$-FSA $\mathcal{A}$ to satisfy $\Gamma = \{1, \ldots, k\} \rightsquigarrow \{k + 1, \ldots, k + l\}$, we then want to generate the (finite) set of outputs $v = \langle v_1 \ldots, v_l \rangle$ for a given input $u = \langle u_1, \ldots, u_k \rangle$. This problem is known to be difficult in general: let $\mathcal{B}$ be a 2-FSA with an unidirectional input tape 1 and a bidirectional output tape 2, and ask if a given input u can produce *any* output v. This problem is equivalent to whether $\mathcal{B}$, considered as a *checking stack automaton*. accepts u [11, Theorem 5.1], which is known to be either **PSPACE**- or **NP**-complete, depending on whether $\mathcal{B}'$ is a part of the instance or not [2, Problem AL5]. However, the additional information Γ provides certain optimization possibilities.

A straightforward way to obtain an evaluation algorithm is to convert the output tapes from *read-only* into *write-once*, and perform these writing operations concurrently with the simulation of the nondeterministic control. Fig. 8 shows the resulting algorithm, where the simulations of all the possible computations are performed in a depth-first order using a stack S. The algorithm maintains for each $1 \leq j \leq l$ an extensible character array $W_j[0 \ldots L_j]$, which holds the contents for the tape squares the output head $k+j$ has already examined during the computation C of $\mathcal{A}$ currently being simulated. Fig. 7 shows the indices during one simulation of the 2-FSA $\mathcal{A}_{\mathrm{rev}}$ from Fig. 1, where the input tape 1 contains the string $\mathbf{ab}$, whose reversal is being generated onto the output tape 2.

In Fig. 8, $T_{\mathcal{A}}$ is enumerated as $\tau_1, \ldots, \tau_{|T_{\mathcal{A}}|}$, and τ_0 is a new starting pseudo-transition $\cdots \xrightarrow[0,\ldots,0]{\cdots} s_{\mathcal{A}}$. We also assume as in Sect. 1.1 that no final state in $F_{\mathcal{A}}$ has any outgoing transitions.

Note that Γ alone does not guarantee that the algorithm in Fig. 8 halts, it just guarantees that only finitely many different outputs are ever generated. Consider namely the situation in Fig. 1, where the 2-FSA $\mathcal{A}_{\mathrm{rev}}$ in Fig. 1 is being used as a transducer in the opposite direction to Fig. 7: input is read from tape 2 and written onto tape 1. As explained in Example 2, $\mathcal{A}_{\mathrm{rev}}$ is now guessing nondeterministically a possible output for later verification against the input. But how long guesses should $\mathcal{A}_{\mathrm{rev}}$ be allowed to make?

This question can be answered by noting that if the currently simulated computation C on input $\boldsymbol{u}$ has written more than

$$B = \mathcal{W}(|\boldsymbol{u}|, \ldots, |\boldsymbol{u}_k|) \tag{3}$$

characters onto some output tape, then C must eventually reject by Definition 2. Hence we can safely add the extra condition $\max(L_1, \ldots, L_l) \leq B + 1$ into branch 15 of the algorithm in Fig. 8 to *prune* all computations C that attempt to generate too long outputs.

Thereafter the stack S will always contain only finitely many different configurations C of the transducer being simulated on input $\boldsymbol{u}$. (These transducer configurations C can be defined in a straightforward by extending the acceptor configurations defined in Sec. 1.1 with write-once output tapes.) Although S represents these configurations C only implicitly, they can be recostructed as in branch 10 of the algorithm in Fig. 8. However, some of these configurations C can still repeat, because the transducer being simulated can also loop on the already known parts of its tapes without generating new output. Fortunately this looping can be detected and eliminated simply by testing in branch 15 of the algorithm in Fig. 8 that the new configuration C_{new} being pushed into S

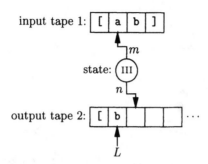

Fig. 7. Simulating the 2-FSA in Fig. 1 as a Transducer.

procedure simulate($\mathcal{A}$:$(k+l)$-FSA;
                    $u_1,\dots,u_k$:Σ^*);

1: Set $m_i \leftarrow 0$ for all $1 \le i \le k$;
2: Set $n_j \leftarrow 0$, $L_j \leftarrow 0$ and $W_j[L_j] \leftarrow$ [for all $1 \le j \le l$;
3: Set $t \leftarrow 1$;
4: Initialize stack S to contain $\langle 0; L_1,\dots,L_l \rangle$ only;
5: **while** S is nonempty **do**
6:    Let $\langle t'; L_1',\dots,L_l' \rangle$ be the top element in S, and let $\tau_{t'} = p \xrightarrow[d_1',\dots,d_k',e_1',\dots,e_l']{\cdots} q$;
7:    **if** $q \in F_\mathcal{A}$ **then**
8:       output$(v_1,\dots,v_l)$, where each $v_j = W_j[1]\dots W_j[L_j-1]$;
9:       Set $t \leftarrow |T_\mathcal{A}| + 1$
10:   **else if** $t > |T_\mathcal{A}|$ **then**
11:      Set $m_i \leftarrow m_i - d_j'$ for all $1 \le i \le k$;
12:      Set $n_j \leftarrow n_j - e_j'$ and $L_j \leftarrow L_j'$ for all $1 \le j \le l$;
13:      Pop off the top element from S;
14:      Set $t \leftarrow t' + 1$;
15:   **else if** $\tau_t = q \xrightarrow[d_1,\dots,d_k,e_1,\dots,e_l]{u_1[m_1],\dots,u_k[m_k],c_1,\dots,c_l} r$ satisfies $(n_j \le L_j) \Rightarrow (W_j[n_j] = c_j)$ for
      all $1 \le j \le l$ **then**
16:      Push $\langle t; L_1,\dots,L_l \rangle$ into S;
17:      Set $m_i \leftarrow m_i + d_i$ for all $1 \le i \le k$;
18:      Set $L_j \leftarrow L_j + 1$ and $W_j[L_j] \leftarrow c_j$ for all those $1 \le j \le l$, on which $n_j > L_j$
      holds;
19:      Set $n_j \leftarrow n_j + e_j$ for all $1 \le j \le l$;
20:      Set $t \leftarrow 1$
21:   **else**
22:      Set $t \leftarrow t + 1$
23:   **end if**
24: **end while**
end simulate;

Fig. 8. An Algorithm for Using Acceptors as Transducers

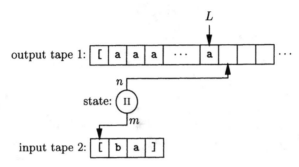

Fig. 9. Generating an Indefinitely Long Output with the 2-FSA in Fig. 1.

does not yet occur in S. This is a standard way to avoid repetition during a depth-first search [13, Chapter 3.6]. We have also experimented with comparing C_{new} against *all* the configurations C encountered so far in the entire search conducted by the algorithm in Fig. 8 on the current input u, but this proved to be extremely inefficient in practice.

Now we have solved problem 2 by developing a halting variant of the evaluation algorithm in Fig. 8. However, this solution suffers from two drawbacks:

1. The cut-off value value B in Eq. (3) is needed to estimate — and hopefully tightly — the pruning depth.
2. The whole stack S must be scanned against repeating configurations C when pushing each new configuration C_{new}.

Fortunately these drawbacks can be alleviated by considering *how* Γ was inferred to hold.

Let Γ be inferred by having the algorithm in Fig. 5 return $\mathbf{1}$ on $\mathcal{A}$ and Γ. Claim 2 in the proof of Theorem 1 shows that every computation $\mathcal{C}$ of $\mathcal{A}$ is "self-limiting" in the sense that no L_j can grow indefinitely. Thus the bound B in Eq. (3) is *not* needed after all, thereby alleviating drawback 1.

Claim 2 alleviates also drawback 2. Two occurrences C_x and C_y of the same configuration during $\mathcal{C}$ have the same color by definition. The proof of the claim shows that C_x and C_y can only arise by traversing a closed loop $\mathcal{L}$, which is not deleted during the algorithm in Fig. 4. We therefore modify this algorithm to *mark* in $\mathcal{A}$ the transitions it considers deleted. Then the algorithm in Fig. 8 can stop scanning its stack S as soon as the most recent marked transition is seen. (In addition, these $\mathcal{L}$ do not advance any output tape $k+j$ into direction $+1$, and therefore this marking technique could be extended even into those transitions in $\mathcal{L}$, which move an output tape into the opposite direction -1 instead.)

Example 3. Applying this modified algorithm in Fig. 4 into the 2-FSA $\mathcal{A}_{\text{rev}}$ in Fig. 1 and $\Gamma = \{1\} \rightsquigarrow \{2\}$ in this way marks every transition by Example 2. Hence the algorithm in Fig. 8 suffices *unmodified* in this case: a compile-time check showed that possibly expensive run-time loop checking is in fact unnecessary, leading into an efficient simulation method for $\mathcal{A}_{\text{rev}}$.

Note that this marking technique can also speed up the simulation of those m-FSAs $\mathcal{B}$, which are still used as acceptors and not transducers: compute the marking given by the modified algorithm in Fig. 4 with $\mathcal{B}$ and $\{1, \ldots, m\}$ (which yields **1**), and use the resulting stack scanning optimization strategy during the simulation of $\mathcal{B}$ on any given input $\langle u_1, \ldots, u_m \rangle$ to identify transitions, which cannot take part in any loops, where a configuration repeats.

A related strategy works also in the case of Theorem 2. Its proof shows that a halting but still correct variant of the algorithm in Fig. 8 can be obtained by adding into its branch 15 the extra condition that configurations C_x and C_y of the same color may not repeat within any computation C: if the path $\mathcal{L}$ of transitions from C_x into C_y advanced any of the unidirectional output tapes $p + q + 1, \ldots, p + q + r$, then this C must be rejecting, while otherwise taking $\mathcal{L}$ during C was unnecessary. Then a variant of the algorithm in Fig. 4, which attempts to mark every transition it can (instead of trying to test for Γ and fail), identifies those cycles $\mathcal{L}$ that can cause some color to repeat. Therefore the same stack scanning optimization strategy as above still applies.

4   Conclusions

We studied the problem of using a given nondeterministic two-way multi-tape acceptor as a transducer by supplying inputs onto only some of its tapes, and asking it to generate the rest. We developed an algorithm for ensuring that this transduction does always yield finite answers, and another algorithm for actually computing these answers when they are guaranteed to exist. In addition, these two algorithms provided a way to optimize the simulation of nondeterministic two-way multi-tape acceptors by restricting the amount of work that must be performed during run-time loop checking.

These algorithms have been implemented in the prototype string database management system being developed at the Department of Computer Science in the University of Helsinki.

Acknowledgments. The author would like to thank Prof. Esko Ukkonen, Prof. Gösta Grahne, Mr. Raul Hakli and Mrs. Hellis Tamm for fruitful cooperation, and Mr. Mika Kekkonen, Mr. Juha Koskelainen, Mr. Jyrki Niemi, Mr. Tero Tuononen, Mr. Ari Vihervaara and Mr. Jaakko Vuolasto for their involvement in implementing the prototype.

References

1. ABITEBOUL, S., HULL, R., AND VIANU, V. *Foundations of Databases.* Addison-Wesley, 1995.
2. GAREY, M., AND JOHNSON, D. *Computers and Intractability: A Guide to the Theory of NP-Completeness.* Freeman, 1979.

3. GINSBURG, S., AND WANG, X. Pattern matching by rs-operations: Towards a unified approach to querying sequenced data (extended abstract). In *ACM SIGACT-SIGMOD-SIGART Symposium on Principles of Database Systems* (1992), pp. 293–300.

4. GRAHNE, G., AND NYKÄNEN, M. Safety, translation and evaluation of Alignment Calculus. In *Advances in Databases and Information Systems* (1997), Springer Electronic Workshops in Computing, pp. 295–304.

5. GRAHNE, G., NYKÄNEN, M., AND UKKONEN, E. Reasoning about strings in databases. In *ACM SIGACT-SIGMOD-SIGART Symposium on Principles of Database Systems* (1994), pp. 303–312.

6. HARRISON, M., AND IBARRA, O. Multi-tape and multi-head pushdown automata. *Information and Control 13* (1968), 433–470.

7. HOPCROFT, J., AND ULLMAN, J. *Introduction to Automata Theory, Languages, and Computation.* Addison-Wesley, 1979.

8. KRISHNAMURTHY, R., RAMAKRISHNAN, R., AND SHMUELI, O. A framework for testing safety and effective computability. *Journal of Computer and System Sciences 52* (1996), 100–124.

9. MECCA, G., AND BONNER, A. Sequences, Datalog and transducers. In *ACM SIGACT-SIGMOD-SIGART Symposium on Principles of Database Systems* (1995), pp. 23–35.

10. NYKÄNEN, M. *Querying String Databases with Modal Logic.* PhD thesis, Department of Computer Science, University of Helsinki, Finland, 1997.

11. RAJLICH, V. Absolutely parallel grammars and two-way finite-state transducers. *Journal of Computer and System Sciences 6* (1972), 324–342.

12. RAMAKHRISNAN, R., BANCILHON, F., AND SILBERSCHATZ, A. Safety of recursive Horn clauses with infinite relations (extended abstract). In *ACM SIGACT-SIGMOD-SIGART Symposium on Principles of Database Systems* (1987), pp. 328–339.

13. RUSSELL, S., AND NORVIG, P. *Artificial Intelligence: a Modern Approach.* Prentice-Hall, 1995.

14. VAN GELDER, A. Deriving constraints among argument sizes in logic programs. Tech. Rep. UCSC-CRL-89-41, University of Califormia, Santa Cruz, 1989. Appears also in the *Annals of Mathematics and Artificial Intelligence.*

15. WEBER, A. On the valuedness of finite transducers. *Acta Informatica 27* (1990), 749–780.

16. WEBER, A. On the lengths of values in a finite transducer. *Acta Informatica 29* (1992), 663–687.

17. WOLPER, P. Temporal logic can be more expressive. *Information and Control 56* (1983), 72–99.

Proving Sequential Function Chart Programs Using Automata

Dominique L'Her, Philippe Le Parc, and Lionel Marcé

LIMI
Université de Bretagne Occidentale
6 av. V. Le Gorgeu, BP 809, 29285 Brest cedex, France
{lher,leparc,marce}@univ-brest.fr

Abstract. Applications described by Sequential Function Chart (SFC) often being critical, we have studied the possibilities of program checking. In particular, physical time can be handled by SFC programs using temporisations, that's why we are interested in the quantitative temporal properties. We have proposed a modeling of SFC in timed automata, a formalism which takes time into account. In this modeling, we use the physical constraints of the environment. Verification of properties can be carried out using the model-checker Kronos. We apply this method to SFC programs of average size like the one of the controlling part of the production cell Korso. The size of the programs remaining however a limit, we are studying the means of solving this problem.

1 Introduction

The language of control in which we are interested is Sequential Function Chart (SFC is the English name of Grafcet). Developed since 1977, this graphical language is based on the step-transition model. Through temporisations, it makes it possible to take time into account. This intuitive and practical language has shown its perfect adaptation to the programming of automated systems. It's one of the languages defined by the IEC1133−3 to program Programmable Logic Controllers. For these, safety is needed ; it is necessary to make sure that specifications are respected by the program. To carry out these verifications, SFC has been modeled in various formalisms which have verification tools.

These modelings have limits: time is not taken into account. However time plays an important role in the command of many automated systems (for instance the timeouts). Thus it is important to treat it. This is why we are interested in temporized SFC and in its temporal verification.

After having presented the main principles of SFC, we will justify the choice of the timed automata for the modeling of SFC. Then this modeling will be described as well as the verifications which it makes possible. Finally we will indicate how we take into account the constraints of the physical world and how the size of the automata can be reduced.

J. M. Champarnaud, D. Maurel, D. Ziadi (Eds.): WIA'98, LNCS 1660, pp. 149–162, 1999.

2   SFC

SFC [2] is a chart model of the behaviour of the control parts of an automated system.

2.1   Structure

The basic graphical elements are (see Figure 1):

- the *steps* which represent the various states of a system. They are symbolized by squares. The initial steps are represented by double squares. During the evolution of an SFC program, the steps are either active or inactive ; during initialization, only the initial steps are active. The set of the active steps of an SFC program at a given moment defines the situation of this SFC program;
- the *transitions* which are used to control the change from a state to another one. They are represented by a horizontal line and control the evolution from step to step. They have two values ; they can be validated or not validated. A transition is validated when all the steps preceding it are active. With any transition, a receptivity is associated, i.e. a Boolean function of the inputs and internal variables of the SFC program like step variables which test if a step is active or not. If a transition is validated and its receptivity has a true value, then this transition is fireable.

  In the receptivities, a particular function renders it possible to measure time: temporisation. The temporisation $t_1/X_i/t_2$ indicates a Boolean condition which takes the value of true if the step i remains active at least t_1 units of time and which will become false t_2 units of time after the deactivation of step i. No structural relation is imposed between the use of temporisation and the referred step i (in the temporisation t_1/X_i, the value of t_2 is implicitly 0).

2.2   Behaviour

Two postulates define the conceptual framework in which SFC must evolve:

- **Postulate 1**: All the events are taken into account as soon as they occur and for all their incidences.
- **Postulate 2**: In the SFC model, causality is considered at null time.

From these two statements, it should be noted that SFC model is sensitive to any external event, whatever its time of occurrence. Any change in the external environment must be taken into account, and the reaction induced must be calculated at null time.

The five following rules define the evolution of an SFC program.

- **Rule 1**: In the beginning, only the initial steps are active.
- **Rule 2**: A transition is validated if all the preceding steps are active. A transition is fireable when it is enabled and its receptivity is true.

- **Rule 3**: A fireable transition is immediately fired. The immediately following steps are then activated and the immediately preceding steps are deactivated. Activations and deactivations are performed simultaneously.
- **Rule 4**: If, in an SFC program, several transitions are simultaneously fireable, they are fired simultaneously.
- **Rule 5**: If a step is simultaneously activated and deactivated, it remains active. The priority is given to activation.

2.3   Interpretation

The behaviour of an SFC program is described by the five rules of evolution. Those are supplemented by interpretation algorithms. The main interpretations are named No Search for Stability (NSS) and Search for Stability (SS).

- **NSS Interpretation**: In the case of the NSS interpretation, a step of evolution corresponds to a simple evolution, that is, the simultaneous firing of all the fireable transitions. Carrying out a step of simple evolution corresponds to the acquisition of inputs, to the computation of the new situation and its output towards the external world.
- **SS Interpretation**: In the case of the SS interpretation, a step of evolution corresponds to an iterated evolution, that is, a simple evolution with acquisition of the inputs followed by a continuation, possibly empty, of simple evolutions without acquisition of inputs, until obtaining a stable situation. A situation is stable when no transition is fireable without taking new inputs. A cycle of instability is a sequence of simple evolutions not leading to a stable situation.

Despite rules and interpretations, ambiguities still persist in the description of SFC programs. For the following modelings, the choices which were made, are detailed in [11].

The SFC program of Figure 1 will illustrate the various points of our talk. At the beginning, steps 0 and 10 are active and it is supposed that input A is false. Transitions 0 and 2 are thus validated but not fireable. Three evolutions are then possible:

- input A becomes true before 10 units of time. In this case, transition 0 is fireable. Its firing causes the deactivation of step 0 and the activation of step 1. The situation { 1,10 } is reached, applying rule 3.
- 10 units of time run out without A becoming true. Transition 2 is fired, the situation { 0,11,12 } is reached, applying rule 3.
- input A becomes true exactly 10 units of time after the activation of step 10. Transitions 0 and 2 are simultaneously fired. The situation reached is { 1,11,12 }, applying rules 3 and 4.

The SFC program then continues to evolve from the current situation.

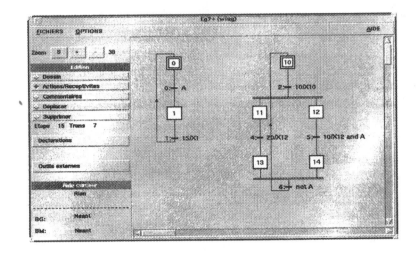

Fig. 1. An SFC program

3 Checking by Using Timed Automata

"Synchronous" languages have been proposed to answer the problems of safe programming. The basic assumption of these languages stipulates that the outputs be considered simultaneously with the inputs that generate them. The SFC language also makes this assumption. In the case of the languages Signal [8] and Lustre [6], the data flow approach still accentuates the proximity between SFC and these languages. On the other hand, if the definition of SFC is purely textual and does not provide a clear semantics, languages Signal and Lustre have a mathematical semantics. Therefore the modeling of SFC in such languages [9] gives us a means of clarifying the semantic choices for SFC. This also allows us to build a simulator and to check properties. However, for checking quantitative temporal properties, this approach is not suitable. Indeed, the discrete representation of time induces an explosion of the number of states of the graph representing all possible runs, making the verifications impossible quickly.

In order to solve this problem of explosion, [5] proposes an approximative method based on convex polyhedrons. Verifications are not performed on the whole graph but on an abstraction.

We have chosen another approach which takes physical time into account. Thus we have studied timed Petri nets [4], timed transition models (TTM) [13], timed automata [7] [12] and hybrid systems [1]. Timed automata give us a good compromise between the power of expression and the possibility of verification.

3.1 Timed Automata

Informally, timed automata are automata extended by a set of real variables, called clocks, the values of which grow uniformly with the passage of time and which can be set to zero. Constraints relating to these clocks are associated with the states and transitions.These timing constraints define the time during which the system can remain in a state and the possibility of firing a transition. Timed automata thus allow a compact modeling of time. Moreover, verifications using model-checking are possible on timed automata. This is why we have chosen them to model SFC.

Definition A timed automaton is a tuple (S, s_{init}, H, A, Inv) where

- S is a finite set of control locations where s_{init} is the initial location.
- H is a finite set of clocks, real variables taking their values in the set of positive real numbers.
- A is a finite set of edges. Each edge is defined by a tuple $(s_s,\ \psi,\ l,\ R,\ s_b)$. s_s and s_b are the source and target locations respectively of the edge. ψ is a timing constraint. To fire the edge, the clocks must satisfy it. l is a label. R is the set of the clocks to be set to zero when the edge is fired. The edge $(s_s,\ \psi,\ l,\ R,\ s_b)$ is also noted $s_s \xrightarrow{\psi, l, R} s_b$.
- $Inv: S \to \psi(H)$ associates with each location a time-progress condition called invariant. While the clocks satisfy the invariant, the system may stay in the location.

At the beginning, the system is at the initial location with all the clocks having the value 0.

Semantics The timed automaton semantics is given by a transitions system $<\mathcal{Q}, \to, (s_0, v_0)>$ where $\mathcal{Q}$ is the set of the states, $\to$ the set of the transitions and (s_0, v_0) the initial state.

- A state (s, v) is a location s and a valuation v of all the clocks.
- The initial state is the pair (s_0, v_0) where s_0 is the initial location and v_0 is the valuation which associates 0 with all the clocks.
- From the state (s_s, v), the transition $(s_s,\ \phi,\ l,\ R,\ s_b)$ can be fired if the valuation of clocks satisfies ϕ. We note $v[R]$ the valuation of the clocks after the firing of the transition which associates 0 with the clocks in R. The clocks in R are set to zero, the values of others clocks remain unchanged. The following rule expresses that :

$$rule\,1: \frac{s \xrightarrow{\phi, a, R} s' \wedge \phi(v)}{(s, v) \xrightarrow{a} (s', v[R])}$$

While the constraint associated to the state is true, the system may stay in the state.

$$rule\,2: \frac{\forall t \in [v, v + d] Inv(s, t)}{(s, v) \xrightarrow{d} (s, v + d)}$$

At any state, the system can evolve either by a discrete state change corresponding to a move through an edge that may change the location and reset some of the clocks, or by a continuous state change due to the progress of time at a location.

3.2   Modeling

First we present the modeling [11] in a general way. Then we specify what each element of a timed automaton represents. A location represents an SFC situation, a value of the inputs and of the temporisations. The transitions correspond to a change of the inputs or an evolution of time inducing a change of the temporisation values. If these changes bring about the firing of some transitions in the SFC program, the target location represents the situation after evolution. The invariants of the states and the temporal constraints express the constraints resulting from temporisations.

Location   In the general case, a location of a timed automaton is defined by a situation of the SFC program, a valuation of the Boolean input variables and the values of the temporisations appearing in the SFC program. Several locations of the automaton can correspond to a single situation of the SFC program.

For the SFC program of Figure 1, if we suppose that input A is false at the beginning, the initial location is $\{0, 10, \overline{A}, \overline{tempo_1}, \overline{tempo_2}, \overline{tempo_3}, \overline{tempo_4}\}$ where $tempo_1$, $tempo_2$, $tempo_3$ and $tempo_4$ indicate the temporisations $10/X10$, $15/X1$, $20/X12$ and $10/X12$ respectively.

Clock   A clock is associated with each step appearing in a temporisation. These clocks have for value, the time since when the associated step has been active or inactive.

For the SFC program of the example, 3 clocks are defined: h_{10} for the step 10 appearing in $tempo_1$, h_1 for the step 1 appearing in $tempo_2$, and h_{12} for the step 12 appearing in $tempo_3$ and in $tempo_4$.

Invariant Associated with a Location   The invariant associated with a location expresses the constraint which the clocks have to satisfy, so that no temporisation changes its value in the location. First of all, we look for the relevant clocks in a location, i.e. those associated with a step, being referred in a temporisation which may change values. They correspond to the clocks which satisfy one of the two conditions:

- the clock is associated with step i, step i is active and there is a false temporisation referring to step i. This temporisation may become true.
- the clock is associated with step i, step i is inactive and there is a true temporisation referring to step i. This temporisation may become false.

The constraint associated with a clock checking the first condition is: $h_i \leq min_j \, t_{1j}$ for $\{\ tempo_j = t_{1j}/X_i/t_{2j}$ with $tempo_j$ false $\}$. The constraint associated with a clock satisfying the second condition is written: $h_i \leq min_j \, t_{2j}$ for $\{\ tempo_j = t_{1j}/X_i/t_{2j}$ with $tempo_j$ true $\}$.

Finally, the constraint associated with a location is true if the location does not comprise any relevant clock or, is the conjunction of the constraints associated with the relevant clocks in the other case.

For example in the initial node, only the clock $h1$ is relevant because only the temporisation $tempo_1$ may change. The invariant is written $h_{10} \leq 10$.

Transition The edge of the timed automata corresponds to a change of the inputs and/or an evolution of time bringing a modification of the values of temporisations. An input may change in any location. Only temporisations corresponding to the relevant clocks in the location may change.

- The **timing constraint** associated with a transition indicates if one or more temporisations change. When the source location does not include any relevant clock, the transition is not constrained temporally: its timing constraint is "true". On the other hand if the source location includes one or more relevant clocks, then the timing constraint is a conjunction of propositions $h_i = t_i$ and $h_j < t_j$. The first form corresponds to a change of temporisation while the second indicates that temporisation remains unchanged.
- The clocks of which steps were activated or deactivated during the transition, are **set to zero** in such a way that the value of the clock, is always the time since when the step has been active or respectively inactive. For example, h_1 is set to zero on the first transition seen Figure 2 because the step 1 is activated.
- From a situation, inputs, edges of inputs, temporisations and edges of steps, the new situation is obtained by a simultaneous firing of all the fireable transitions and the new value of temporisations is computed. The **target location** is then defined by the situation reached, the inputs and temporisations kept up to date.

The transitions of the timed automaton do not inevitably correspond to the firing of SFC program transitions.

From the initial location $\{0, 10, \overline{A}, \overline{tempo_1}, \overline{tempo_2, tempo_3, tempo_4}\}$, three transitions are possible according to whether the input A and/or the temporisation $tempo_1$ become true. In Figure 2, these transitions are described.

Construction The construction of the timed automaton starts with the definition of the initial location. Then this location is treated, i.e. its invariant and the transitions leaving it are computed. Then new locations are in general built. The construction continues, as long as not all the locations have been treated. The number of possible locations being finite the algorithm ends.

The complete automaton representing the SFC program of the Figure 1 has 3 clocks, 14 locations and 58 transitions in the case of the SS interpretation.

3.3 Checking

We model the SFC program with a timed automaton in order to check properties using the verification tool Kronos [3].

The properties expressed on the SFC level, must be translated into Timed Computation Tree Logic (TCTL) to be checked by the model-checker Kronos.

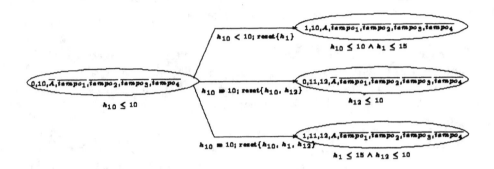

Fig. 2. First step of the construction of the timed automaton for the SFC program of Fig. 1

This translation is an important stage of the checking. It requires a thorough knowledge of logic and often requires to express the respective property very precisely.

TCTL   TCTL [7] is a temporal logic which extends arborescent logic CTL by introducing a global variable: time. As a tree logic, TCTL uses symbols that concern at the same time the set of all possible executions ($\exists$: there is an execution, $\forall$: for all the executions) and the set of execution states ($\Diamond$: there is a state, $\Box$: for all the states, $\mathcal{U}$: until a state). In order to introduce time explicitly into the syntax, the scope of the temporal operators is time-limited. Thus, the formula $\forall\Box_{\leq 4}p$ intuitively means that, for all the executions of the system, proposition p is true for all the states until the fourth time unit.

Some Properties TCTL, although reserved to express quantitative temporal properties, renders it possible to write the usual qualitative temporal properties.

Thus to check that a situation is a deadlock, i.e. a situation which it is not possible any more to leave, various formulae can be defined. The following formula makes it possible to know if the situation S is reachable and is always a deadlock: $(init \Rightarrow \exists\Diamond\ S) \land (init \Rightarrow \forall\Box\ (S \Rightarrow \forall\Box\ S))$.

Without being always a deadlock, a situation may be locked in some cases: $init \Rightarrow \exists\Diamond\ (S \land (S \Rightarrow \forall\Box\ S))$.

We show that the situation $\{0,11,14\}$ is not a deadlock but on the other hand, once the steps 11 and 14 are reached, they infinitely remain active.

On timed automata, we can also check quantitative temporal properties:

- We can check the duration of activation of a step: does step i remain active more than (at least) t units of time? For instance, we check that step 1 could remain active more than 15 units of time by showing that the following formula is true:

init $\Rightarrow \exists \Diamond$ (e0_1 $\wedge$ (e0_1 $\Rightarrow$ e0_1 $\exists \mathcal{U}_{>15}$ e0_1)) where e0_1 is the proposition associated with the location when the step 1 is active.

- We can also study the time which separates two activations of distinct situations S_1 and S_2. Thus to show that between the activation of S_1 and the activation of S_2, the maximum duration is lower than t, the following formula must be false:

$init \Rightarrow \exists \Diamond (S_1 \wedge (S_1 \Rightarrow (\neg S_2 \exists \mathcal{U}_{>t} \neg S_2)))$.

Using the Kronos tool, we succeed in checking properties on the timed automata resulting from the SFC programs. By this method, we can check SFC programs of more important size and which have more temporisations. Moreover delays are not a limitation any more. Indeed, the complexity of the algorithm of verification is independent of delays.

4   Applications

We have studied the production cell Korso [10]. The programming of the control of this application is achieved very easily in an SFC program. For the checking, we have to solve two problems: taking environment into account and reducing size of the automata. We present the solutions we have found and the tools we have developped.

4.1   Taking Environment into Account

To explain the problem, we take an element of the operative part of Korso, the press as an example. The press consists of a horizontal plate which can move vertically. The SFC program of the press is given in Figure 3. In steps 50 and 51, the press waits in the median position (*cap2*) until a metal blank is loaded by the robot (step 33 of the robot is then reached). Then it goes up (step 52) to the high position (*cap3*) and the metal blank is worked (step 53) during 2 units of time. Then the press goes down (step 54) to the low position (*cap1*) where it waits (step 55) to be discharged by the robot (step 41 or 42 or 43 of the robot). Finally it goes up (steps 56 and 57) until reaching its median initial position again. The process can then start again.

This SFC program is synchronized with the robot by the step variables X33, X41, X42 and X43. To study it separately, we let go these synchronizations by replacing "X33" by the variable vX33 and " X41 or X42 or X43 " by the variable vX4. These variables vX33 and vX4 evolve freely.

We build the timed corresponding automaton. It has 1296 locations, 262 896 transitions and 3 clocks. It is too large to be checked by the Kronos tool which accepts only automata having fewer than the 65000 transitions.

Moreover, the automaton has locations which represent the press in the high and low positions simultaneously. In normal running, these have no sense ; they

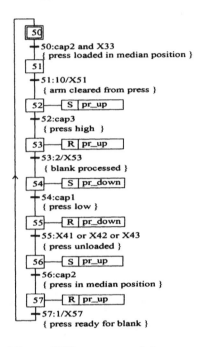

Fig. 3. SFC program of the press

do not fulfill the constraints of the environment. This is why the construction of the automaton was then modified so that only the locations satisfying the constraints of the environment are considered. During our study, we have encountered three kinds of constraints, according to how they relate to the locations and/or the transitions:

– Only one of the sensors $cap1$, $cap2$, $cap3$ may be true at one moment because the press is in a single position. A stronger constraint can be expressed if inputs $c12$ and $c23$ (representing the position between top and medium and the position between medium and low) are introduced. In this case, it is necessary that there should be one and only one of the sensors ($cap1$, $c12$, $cap2$, $c23$, $cap3$) true at a given moment. The locations which do not check this constraint are removed.

– The changes of values of the sensors are constrained to pass from the low position to the high position by a medium position. The transitions which do not satisfy this constraint are removed.

– The constraints handling, at the same time, the locations and the transitions express the links which exist between the actions and the sensors. Thus when the action pr_up is done, the low position is not more reachable. In the same way when the action pr_down proceeds, reaching the high position is not possible any more.

We are interested in two properties of the press:

- The formula expressing that the press should not be moved to low if the sensor $cap1$ is true, is written: init $\Rightarrow \forall\square \neg$ (pr_down and cap1)
- In the same way, to show that the press should not be moved to high if the sensor $cap3$ is true, it should be shown that the following formula is true: init $\Rightarrow \forall\square \neg$ (pr_up and cap3)

By introducing the inputs $c12$ and $c23$ and by considering only the first two kinds of constraints, the automaton built has 922 locations and 57606 transitions. On this automaton, the two properties are false.

Moreover while inserting, the constraints resulting from the actions, the automaton then has 314 locations and 13670 transitions.

The first property is always false, this is due to the relaxation of synchronizations which produces an instability. Thus step 55 is not always activated ; it follows that the action *"stop to go down"* is not always carried out when $cap1$ is true.

The second property is true showing that the environment has been taken into account sufficiently.

Working on a more realistic representation, we can check more properties. Taking into account the environment makes it possible to decrease the size of the automaton but does not solve all the problems of size.

4.2 Reduction of the Size

During our verifications, we wish to know which situations are reachable and which values can take the inputs in these situations. The given modeling makes it possible to answer these questions. It is however possible to consider other modelings solving this problem. If a Boolean formula could be associated with a location, the most compact modeling would consist of a timed automaton reduced to the graph of the situations. As only conjunctions of the propositional variables can be associated with the locations, we propose a smaller modeling than initial modeling but not reaching the graph of the situations.

In a location, we do not indicate the value of each input any more but only the value of the important inputs. For a particular situation, an input is important if a modification of its value can induce an evolution of the SFC program. In the timed automaton, a transition is defined only if it corresponds to a modification of the important inputs or a modification of temporisations.

The initial situation obtained, we determine the important inputs for this situation. The values of these important inputs are then fixed. The initial location is completely defined. For example, for the grafcet of the Figure 4 and WORS interpretation, the only important input for the step 0 is a. As a is initially false, the initial location is written $(0, \bar{a}, \hat{b}, \hat{c})$ where $\hat{e}$ means that the value of the input e does not have importance for the current situation.

Then, as long as there remains a state to be treated, we construct the whole automaton by the following operations: for each temporal event, for each possible

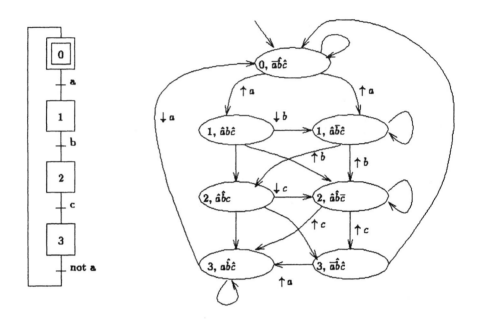

Fig. 4. SFC program and corresponding timed automaton for the new modeling in NSS interpretation (the timing constraints are not written)

combination of the important inputs of this state, we study the target situation. If an input is important for the target location, its value must be fixed. Two cases can occur:

- either it is important in the source location, its value is then perfectly defined. It is the case of the input b for the evolution from $(1, \hat{a}, b, \hat{c})$ corresponding to $\downarrow b$.
- either it is not important in the source location. The target location is divided into two sets of locations, one representing the true input, the other the false input. For example, the virtual location $(2, \hat{a}, b, \hat{c})$ is reachable from the location $(1, \hat{a}, b, \hat{c})$. This location, where c is a important input, is divided into two: $(2, \hat{a}, \hat{b}, c)$ and $(2, \hat{a}, \hat{b}, \bar{c})$.

If an input is not important in the target location, then if it is important in the source location, its value is free. For example, the input b of the virtual location $(2, \hat{a}, b, \hat{c})$ is slackened for example in $(2, \hat{a}, \hat{b}, c)$.

In this modeling, a location represents several locations of the preceding modeling, in the same way the number of transitions is reduced. Thus for the example and NSS interpretation, the timed automaton has 22 locations and 176 transitions in the first modeling and 7 locations and 19 transitions for the new modeling. For SS interpretation, the timed automaton has 16 locations and 112 transitions for the first modeling and 11 location and 66 transitions for the new modeling. The reduction is more sensitive for the NSS interpretation than the SS interpretation ; indeed, the number of significant inputs relative to a situation is smaller in NSS than in SS.

This modeling makes it possible to decrease the size of the timed automata in a considerable way. On the other hand, it is difficult to take environment into account with this modeling. Indeed, as it is not possible to consider Boolean formulas at the level of the locations, the constrained inputs must be considered important in all situations. Therefore in the worst case, that is to say all the inputs are constrained, no profit will be obtained because of new modeling.

We have also studied techniques allowing to decrease the size of the systems to be checked: the composition and the abstraction. These techniques are powerful. However, for the timed systems, their study is relatively recent and few results have been obtained. Their application to the checking of SFC programs is not immediate and still requires basic work on the timed systems.

4.3   Tools

To facilitate the design and the checking, various tools have been built such an editor of SFC progams (see Figure 1), a simulator, the translators SFC programs-timed automata as well as an interface of verification.

The interface makes it possible to choose the parameters of the checking and to execute the chain of tools which produce the result of the checking.

It was developed in Tcl/Tk and it is composed by a control panel (see Figure 5). From this one, the user can choose the various parameters of the verification:

- the SFC program.
- the property which he wants to check.The properties are expressed in a literal way. They are reachable in a tree structure.
- options. The choice of interpretation (SRS or ARS) is possible. We can moreover specify if the simultaneous modifications of inputs are authorized or not. The possibility of taking the environment into account was also given. For each type of constraint defined in the paragraph 4.1, a window of data entry has been defined.

When the checking is started, the interface takes care of several tasks:

- construction of the TCTL formula corresponding to the property,
- construction of the timed automaton corresponding to the SFC program
- call the tool of verification

On the Figure 5, the result of the verification of a property on the SFC is shown.

5   Conclusion

By this work, we show that it is possible to take time into account in modeling of SFC programs and to check their qualitative or quantitative temporal properties.

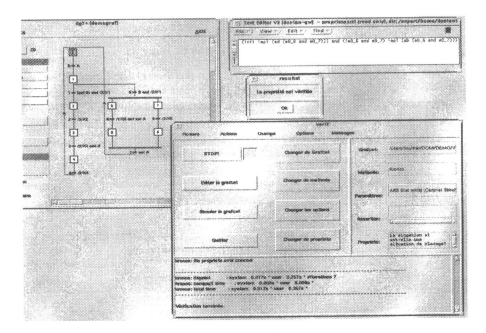

Fig. 5. Interface of verification

In modeling with the synchronous languages, we represent discrete time. During the checking, this modeling leads to a combinatorial explosion of the number of states, each moment being represented by a state.

We then have turned to timed automata. This formalism takes into account continuous time in its definition, owing to real variables called clocks. With each step i referred in a temporisation $t_1/X_i/t_2$, we associate a clock. This computes the time since which the step has been active or inactive. On a timed automaton resulting from this modeling, we could check temporal properties such accessibility in a minimum or maximum time, or the durations of minimum activity and maximum. For this verification, the delays are not a limitation any more.

On the other hand, the size of the automata remains a barrier to the checking. An automaton should not have more than 65000 transitions, so that Kronos can treat it. Unfortunately, some of the automata generated from the SFC programs can have more than 100000 transitions. In order to solve this problem, solutions have been studied. Taking the environment at the level of the states and the transitions into account enables us to decrease the size of the automata considerably. In the same way, a proposed new modeling makes it possible to reduce the number of states and transitions from the automata generated. However to increase the size of the SFC programs that can be treated, efforts must continue in these directions as in the study of the verification techniques of composition and abstraction.

To ensure more safety of SFC programs, checking only does not seem to be sufficient. In parallel, we think that a methodology should be developed making it possible to avoid design errors. This methodology could perhaps also support the building of more easily verifiable SFC programs. It seems also important to us to confront the models and theories already developed with the industrial applications.

We wish to thank the anonymous referee who helped us to improve our pidgin English.

References

1. R. Alur, C. Courcoubetis, N. Halbwachs, T. Henzinger, P. Ho, X. Nicollin, A. Olivero, J. Sifakis, and S. Yovine. The algorithmic analysis of hybrid systems. *Theoretical Computer Science*, 138:3–34, 1994.
2. N. Bouteille, P. Brard, G. Colombari, N. Cotaina, and D. Richet. *Le GRAFCET*. "CÉPADUÈS ÉDITION", Toulouse, France, 1992.
3. C. Daws, A. Olivero, S. Tripakis, and S. Yovine. The tool Kronos. In R. Alur, T. Henzinger, and E.D. Sontag, editors, *DIMACS Workshop on Verification and Control of Hybrid Systems*, pages 208–219. LNCS 1066, Springer-Verlag, 1995.
4. C. Ghezzi, D. Mandrioli, S. Morasca, and M Pezzè. A unified high-level Petri net formalism for time-critical systems. *IEEE Transactions On Software Engineering*, 17(2):160–172, February 1991.
5. N. Halbwachs. Delay analysis in synchronous programs. In C. Courcoubetis, editor, *5th Conference on Computer-Aided Verification*, pages 333–346. LNCS 697, Springer-Verlag, 1993.
6. N. Halbwachs, F. Lagnier, and C. Ratel. Programming and verifying real-time system by means of the synchronous data-flow language Lustre. *IEEE Transactions on Software Engineering*, 18(9):785–793, September 1992.
7. T.A. Henzinger, X. Nicollin, J. Sifakis, and S. Yovine. Symbolic model-checking for real-time systems. *Information and Computation*, 111(2):193–244, 1994.
8. P. Le Guernic, T. Gautier, M. Le Borgne, and C. Le Maire. Programming real time applications with Signal. *Proceedings of the IEEE*, 79(9):1321–1336, 1991.
9. P. Le Parc, D. L'Her, J.L. Scharbarg, and L. Marcé. Le Grafcet revisité à l'aide d'un langage synchrone flot de données. *Technique et Science Informatiques*, 17(1):63–86, January 1998.
10. C. Lewerentz and T. Lindner. Case study 'production cell' : A comparative study in formal specification and verification. In C. Lewerentz and T. Lindner, editors, *Formal Development of Reactive Systems : Case Study Production Cell*, pages 1–54. LNCS 891, Springer-Verlag, 1995.
11. D. L'Her. *Modélisation du Grafcet temporisé et vérification de propriétés temporelles*. PhD thesis, Université de Rennes 1 (FRANCE), September 1997.
12. X. Nicollin, J. Sifakis, and S. Yovine. Compiling real-time specifications into extended automata. *IEEE Transaction on Software Engineering. Special Issue on Real-Time Systems*, 18(9):794–804, 1992.
13. J.S. Ostroff. Automated Verification of Timed Transition Models. In G. Goos and J. Hartmanis, editors, *Workshop on Automatic Verification Methods for Finite State Systems*, pages 247–256. LNCS 407, Springer-Verlag, 1989.

Automata and Computational Probabilities

Marie-Chantal Beaulieu and Anne Bergeron

LACIM
Université du Québec à Montréal
C.P. 8888 Succursale Centre-Ville, Montréal, Québec, Canada, H3C 3P8
{anne, mcbeau}@lacim.uqam.ca

Abstract. In this paper, we discuss the underlying ideas of a computer laboratory for symbolic manipulation of discrete random experiments. Finite automata, and associated formal series, are the basic theoretical tool for representing experiments, and for solving probability problems. Starting from a description of a random experiment given as a special kind of regular expressions, the environment constructs automata from which it extracts generating series associated to the experiment.

1 Introduction

The interaction between discrete probabilities and formal language theory has a long story. Probabilistic tools have been used early in information theory [11], automata theory [5] and coding theory [3], [8]. More recently, evidence of the usefulness of this interaction have been found in fields like asymptotic analysis of algorithms [6], allocation problems, concurrency measures [7], communication protocols specification and verification, and random generation of combinatorial structures.

Most of these approaches rely on describing various subsets of A^*, the set of words on a finite alphabet A, defining probability measures on the elements of these subsets, and then deriving formal properties with the usual constructions of probability theory. We are interested in developing the underlying computational aspects of these methods in order to construct symbolic tools that can be used to assist research or education in these fields.

The main problem can be stated as the *formal manipulation of discrete random experiments*. This primarily involves describing *algebras* of experiments and linking these to algorithms that can compute adequate probabilistic parameters.

2 Discrete Random Experiments

A discrete random experiment is often described by the set Ω of its possible outcomes, and by a function

$$p : \Omega \longrightarrow [0, 1]$$

assigning a probability $p(e)$ to each event $e \in \Omega$, such that $\sum_{e \in \Omega} p(e) = 1$.

J.-M. Champarnaud, D. Maurel, D. Ziadi (Eds.): WIA'98, LNCS 1660, pp. 164–177, 1999.

For example, consider the experiment of tossing a coin repeatedly until either the pattern *head-head-tail* or *head-tail-tail* shows up [P 74]. Writing h for *head* and t for *tail*, the set Ω of possible outcomes is:

$$\Omega = \{hht, htt, hhht, thht, thtt, \ldots\}$$

If we assume that the coin is fair, and that the trials are independent, we can assign to each sequence $r_1 r_2 \ldots r_n$ in Ω the probability $(1/2)^n$.

We are interested in answering the following type of questions about the experiment:

> *What is the probability that the pattern head-head-tail shows up before the pattern head-tail-tail?*
> *What is the mean length of a sequence?*

Such questions can be answered with classical tools of probability theory, but the computations involve infinite series that are often difficult to represent or to sum. Our goal is to automate the representation and solutions to such problems, at least for some classes of random experiments. (Answers to the above questions are given in Section 4.3.)

2.1   Trials and Experiments

In the sequel, we will focus on the class of discrete experiments that are independent repetitions of an elementary trial.

Definition 1 *A trial is a simple random experiment whose set of possible outcomes is finite.*

For example, *tossing a coin* is a trial with possible outcomes $A = \{h, t\}$. To each event in the set of possible outcomes of a trial, we assign a numerical probability:

$$p(h) = 1/3, \quad p(t) = 2/3,$$

or a symbolic one:

$$p(h) = p, \quad p(t) = 1 - p,$$

such that the sum of probabilities is equal to 1.

Definition 2 *An experiment is the independent repetition of a trial until the sequence of results $e = r_1 r_2 \ldots r_n$ exhibits a given property, called the stopping condition.*

By the independence hypothesis, we can assign the probability

$$p(e) = p(r_1)p(r_2) \ldots p(r_n)$$

to an element $e = r_1 r_2 \ldots r_n \in \Omega$, where $p(r_i)$ is the probability that the elementary trial yields r_i.

Note that, in this definition, it is assumed that no proper prefix of $r_1 r_2 \ldots r_n$ satisfies the stopping condition. For example, in the coin tossing example, the sequence *hhtt* does not belong to Ω since the experiment would have been stopped after the third throw.

Furthermore, it is not clear if Definition 2 captures the usual notion of random experiment. Indeed, the concept of *stopping condition* is quite vague. For example, a condition could be hard to test, as in *"pick up random digits and stop when the last ten appear consecutively in the decimal expansion of π"*. Even easy to test but improbable properties such as *"toss a fair coin until the number of heads equals a hundred times the number of tails"* suggest that some experiments are not guaranteed to stop.

In the sequel, we will restrict the possible outcomes of an experiment to *finite* sequences of results, and we will want to have reasonable evidence that one of these outcomes will happen. We will formalize these concepts in the next section.

3   Representing Experiments with Automata

The problem of describing the set of finite possible outcomes Ω associated to an experiment can easily be reformulated as a language theoretic problem: *Let the finite alphabet A be the set of outcomes of a trial, then the set Ω of finite possible outcomes of an experiment is a subset of A^*.*

It is thus natural to turn to automata theory in order to investigate different formalisms for describing sets of possible outcomes. Of these many formalisms, finite automata are the best suited for symbolic computations. We will see that if Ω can be represented by a finite automaton, then most questions about the experiment can be symbolically derived from the automaton.

Consider the example of tossing a coin until either *hht* or *htt* appears. The set Ω of possible outcomes can be represented by the automaton of Fig. 1, with initial state 0, and final state 4.

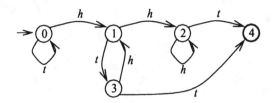

Fig. 1. Automaton recognizing Ω

This automaton is a simple mechanism to test membership to the set Ω. As a generating device, it can also be used to simulate an experiment: in each state, the result of the elementary trial is randomly chosen according to the given probabilities, and the automaton goes to its next state by following the transition corresponding to the result of the trial.

Since our goal is to allow the definition and formal manipulation of random experiments in a computer environment, we must design a formal language that allows the construction of complex experiments starting from simple ones. Consider, for example, the stopping condition:

$$\text{"...until htt or hht appears"}.$$

It generates the set of possible outcomes:

$$\Omega_1 = \{hht, htt, hhht, thht, thtt, \dots\}.$$

The above condition can be naturally split in two simpler conditions:

$$\text{"...until htt appears"},$$
$$\text{"...until hht appears"}.$$

Which, respectively, generates the sets of possible outcomes:

$$\Omega_2 = \{htt, hhtt, thtt, hhhtt, hthtt, tthtt, \dots\},$$
$$\Omega_3 = \{hht, hhht, thht, hhhht, hthht, tthht \dots\}.$$

Since the initial condition is a disjunction, the obvious choice to express Ω_1 would be as the union of Ω_2 and Ω_3, but it does not work. For example, the result $hhhtt$ belongs to $\Omega_2 \cup \Omega_3$, but is not in Ω_1 since it has a proper prefix ending with hht.

Thus, complex experiments cannot be directly constructed from simpler ones using boolean operators. However, in the following section, we will consider a class of languages that is closed under a few basic operations, and that will provide a suitable frame to express random experiments.

3.1   Semaphore Codes

Let A be a finite alphabet, and s be a non empty word in A^*. Consider the experiment with a simple stopping conditions of the form "...until the sequence s appears". The set Ω of possible outcomes of this experiment can be described as the set of sequences that end with s, but for which no proper prefix ends by s. This set can be represented by the regular expression:

$$\Omega = A^* s - A^* s A^+.$$

This expression gives the definition of Ω in a compact form, and can be applied to any non empty word s. Moreover, if s and t are two non empty words in A^*, the set of sequences that end with either s or t, but for which no proper prefix ends by s or t, can be represented by the regular expression:

$$A^*(s + t) - A^*(s + t)A^+.$$

This representation displays nicely where the conjunction *or* of the stopping condition effectively occurs, ie. in $(s + t)$. Sets constructed this way are studied in coding theory under the name of J-codes [10], or *semaphore codes* [3]. We have the following definition:

Definition 3 *Let A be an alphabet, and S be a non-empty subset of A^*, then the set:*

$$X = A^*S - A^*SA^+$$

is called a semaphore code. The set S is the set of semaphores of X.

The most important consequence of this definition is that any formalism used to describe rational sets can be used to describe stopping conditions, as long as it is understood that only the semaphores are described. Indeed, given a rational set S of stopping conditions, the set $A^*S - A^*SA^+$ will also be rational, thus representable by a finite automaton.

3.2   Constructing Automata in the Laboratory

Semaphore codes enable us to construct a simple formalism to describe random experiments with stopping conditions. We chose three basic constructions to which we associate algorithms that construct the automaton recognizing the corresponding set of possible outcomes.

A *basic* experiment on an alphabet A is an experiment with a stopping condition of the form:

"...until the word s appears"

where s is a non empty word of A^*. The notation for a basic experiment will be simply the word s.

In the preceding section, we saw that this set is recognizable by a finite automaton. Indeed, constructing the corresponding can be done with a classical string matching algorithm that uses automata. Given a word $s = s_1 s_2 \ldots s_n$, we first construct a non-deterministic automaton that recognizes all sequences that contains s as a factor (Fig. 2).

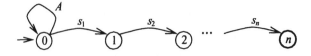

Fig. 2. Non-deterministic automaton recognizing $s = s_1 s_2 \ldots s_n$

It can be shown, see [1] for example, that the corresponding deterministic automaton has also n states, and only one final state. Since we are interested in the first occurrence of the pattern, we remove any transition that is defined in the final state. For example, the stopping condition *"...until hht appears"* yields the automaton of Fig. 3.

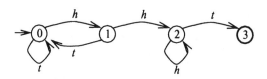

Fig. 3. Automaton recognizing sequences ending with hht

In order to describe more complex experiments, we define two operators. The first one is a disjunction, written as $E_1 + E_2$ and expresses the stopping condition:

"...until either condition E_1 or condition E_2 is satisfied"

For example, the condition *"...until htt or hht appears"* will be written as:

$$hht + htt.$$

The second operator is used to express sequences of conditions. The expression $E_1 * E_2$ will correspond to the stopping condition:

"until condition E_1 is satisfied then condition E_2 is satisfied"

For example, the condition *"...until h appears and then t appears"*, whose set of possible outcomes is:

$$\{ht, tht, hht, hhht, thht, ttht, \dots\}$$

would be written as:

$$h * t.$$

To each expression E constructed with words and the two operators $+$ and $*$, we associate a set $\mathcal{L}(E)$. If E is the word s, then $\mathcal{L}(E)$ is the semaphore set:

$$A^* s - A^* s A^+.$$

The sets associated to complex expressions are sets recognized by automata recursively constructed with the following rules, where $\mathcal{A}_1$ is the automaton associated with E_1, and $\mathcal{A}_2$ the automaton associated with E_2. Each rule yields an automaton that has one final state with no outgoing transition.

1) The automaton $\mathcal{A}$ associated to the stopping condition $E_1 + E_2$ is obtained by computing the Cartesian product of the two automata $\mathcal{A}_1$ and $\mathcal{A}_2$, and by considering any state (s_1, s_2) of the product as final if either s_1 or s_2 is final.

The initial state of $\mathcal{A}$ is (i_1, i_2) where i_1 is the initial state of $\mathcal{A}_1$, and i_2 is the initial state of $\mathcal{A}_2$. The transition

$$(s_1, s_2) \cdot r$$

is defined if and only if $s_1 \cdot r$ and $s_2 \cdot r$ are defined in $\mathcal{A}_1$ and $\mathcal{A}_2$.

If both $\mathcal{A}_1$ and $\mathcal{A}_2$ have only one final state with no outgoing transition, then any final state (s_1, s_2) of $\mathcal{A}$ will have no outgoing transition, thus all these states are equivalent and we can identify them.

2) The automaton $\mathcal{A}$ associated to the stopping condition $E_1 * E_2$ is obtained by gluing the final state of $\mathcal{A}_1$ to the initial state of $\mathcal{A}_2$, as in Fig. 4. The initial state of $\mathcal{A}$ will be the initial state of $\mathcal{A}_1$, and its final state will be the final state of $\mathcal{A}_2$.

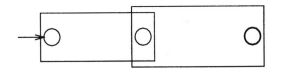

Fig. 4. Gluing $\mathcal{A}_1$ and $\mathcal{A}_2$

If the final state of $\mathcal{A}_1$ has no outgoing transition, then this construction yields a deterministic automaton. If the final of $\mathcal{A}_2$ has no outgoing transition, then $\mathcal{A}$ inherits that property by construction.

Thus, any expression describes a set of sequences that can be recognized by a deterministic automaton. Moreover, all sets constructed this way are semaphore codes. Indeed, basic experiments of the form *"...until the sequence s appears"* are semaphore codes with s as semaphore, and we have:

Proposition 1 *If $\mathcal{L}(E_1)$ and $\mathcal{L}(E_2)$ are semaphore codes, then $\mathcal{L}(E_1 + E_2)$ and $\mathcal{L}(E_1 * E_2)$ are also semaphore codes.*

Proof: Let S_1 and S_2 be the semaphores of $\mathcal{L}(E_1)$ and $\mathcal{L}(E_2)$ respectively. Then the set of semaphores of $\mathcal{L}(E_1 + E_2)$ is simply the union of S_1 and S_2. Indeed, if $e \in \mathcal{L}(E_1 + E_2)$ then, by construction, e is a prefix of words in both $\mathcal{L}(E_1)$ and $\mathcal{L}(E_2)$, thus e has no proper factor that belong to $S_1 \cup S_2$. Since e is also a word of either $\mathcal{L}(E_1)$ and $\mathcal{L}(E_2)$, it ends with a word in $S_1 \cup S_2$.

On the other hand, suppose that $e \in A^*(S_1 + S_2) - A^*(S_1 + S_2)A^+$, and suppose that e can be written as $e's$ where $s \in S_1$. Then e is a word of $\mathcal{L}(E_1)$, and since e does not contain any proper factor in S_2, it is a prefix of a word in $\mathcal{L}(E_2)$. Thus e belongs to $\mathcal{L}(E_1 + E_2)$ by construction.

In order to prove the second part of the proposition, we will show that the set of semaphores of $\mathcal{L}(E_1 * E_2)$ is $S_1 \cdot \mathcal{L}(E_2)$. If $e \in \mathcal{L}(E_1 * E_2)$, then e can be written as $e_1 s_1 e_2 s_2$ with $s_1 \in S_1$ and $s_2 in S_2$, thus e ends with a word in $S_1 \cdot \mathcal{L}(E_2)$. Moreover e cannot have a proper prefix of the form $s'_1 e'_2 s'_2$ since it would imply that $e_1 s_1$ would have s'_1 as a proper factor, or $e_2 s_2$ would have s'_2 as a proper factor.

Suppose now that e belongs to the set with semaphores $S_1 \cdot \mathcal{L}(E_2)$. Then e can be written as $e_1(s_1 e_2 s_2)$, thus it is a prefix of a word in $\mathcal{L}(E_1 * E_2)$. Since e has no proper factor in $S_1 \cdot \mathcal{L}(E_2)$, we have $e \in \mathcal{L}(E_1 * E_2)$. ∎

3.3   Examples and Counterexamples

The following examples show that a vast range of experiments can be defined with the expressions of Section 3.2. In these examples, we suppose that the possible outcomes of a trial are $A = \{a, b, c\}$, and that n is a natural number.

Example 1. Stop after n trials:

$$(a + b + c) * (a + b + c) * \ldots * (a + b + c).$$

Example 2. Stop after n occurrences of the symbol a:

$$a * a * \ldots * a.$$

Example 3. Stop after each symbol of A appears at least once:

$$(a * b * c) + (a * c * b) + (b * a * c) + (b * c * a) + (c * a * b) + (c * b * a).$$

On the other hand, some stopping conditions cannot be represented by such expressions. The simplest counterexample is to consider a two symbol alphabet $A = \{a, b\}$ with the stopping condition "*until the number of a is equal to the number of b*". It is well known in automata theory that the corresponding set Ω is not rational.

Finally, even if semaphore codes can express a wide variety of experiments, not all experiments are semaphore. Consider, for example, the following set of sequences over the alphabet $\{a, b\}$:

$$\{a, baa, bab, bb\}$$

Suppose that one defines an experiment that stops when the sequence of all trials is equal to one of those four sequences. If we set $p(a) = 1/2$ and $p(b) = 1/2$, then, assuming independence, the sum of the probabilities of the four sequences in this set is equal to 1. Thus the experiment stops after at most 3 trials but, since a is a proper factor of baa, this set is not semaphore.

3.4   Does It Stop?

We now turn to the problem of deciding if an experiment will stop or, to put this more formally: *Does the sum of the probabilities of all events in Ω is equal to 1?*

This is clearly not true for most subsets of A^*, but one can find many positive results within code theory. All automata constructed with the notation of the previous section are semaphore codes, and one of the nice aspect of using semaphore codes is that the corresponding random experiment will always stop if the probabilities associated to the results of a trial are positive. The following result is a consequence of a more general theorem that can be found in [HP 89].

Proposition 2 *Let A be a finite set and $p : A \to [0,1]$ such that $p(r) > 0$ for all $r \in A$, and $\sum_{r \in A} p(r) = 1$. The function p can be extended multiplicatively to words in A^*.*

Let $S \in A^*$ be a non empty rational set, then if $\Omega = A^* S - A^* S A^+$ we have

$$\sum_{e \in \Omega} p(e) = 1$$

∎

Thus, any experiment whose stopping condition is described with the operators of Section 3.2 will stop with probability 1.

4   Symbolic Computations

Once the possible outcomes of an experiment is represented by a finite automaton, it is possible to answer algorithmically to many questions about the experiment. In this section, we give the flavor of these algorithms whose details can be found in [B 98].

4.1   Series Associated to an Experiment

Let's consider again the coin tossing example. A way to encapsulate the information about the experiment is to write it as an - infinite - formal sum:

$$E(h,t) = \frac{1}{8} hht + \frac{1}{8} htt + \frac{1}{16} hhht + \frac{1}{16} thht + \frac{1}{16} thtt + \frac{1}{32} hhhht + \cdots$$

This sum is obtained by formally multiplying a result e of the experiment by its probability $p(e)$, and summing over all possible values $e \in \Omega$, that is:

$$E(h,t) = \sum_{e \in \Omega} p(e)e$$

This function can be used to answer questions about some random variables defined over the experiment. For example, consider the random variable associating to a result its length, then the generating function $F(x)$ of this variable is given by:

$$F(x) = E(x,x) = \frac{1}{8}x^3 + \frac{1}{8}x^3 + \frac{1}{16}x^4 + \frac{1}{16}x^4 + \frac{1}{16}x^4 + \frac{1}{32}x^5 + \cdots$$

By differentiating this function, and evaluating the result for $x = 1$, we obtain the mean length of the experiment as the series:

$$F'(x)_{x=1} = \frac{1}{8}3 + \frac{1}{8}3 + \frac{1}{16}4 + \frac{1}{16}4 + \frac{1}{16}4 + \frac{1}{32}5 + \cdots$$

All these observations make little sense if there is no way to get rid of the infinite series. As we will show in the next section, for the class of experiments that can be described in the laboratory, it is always possible to obtain a closed form for the expression:

$$E(h,t) = \sum_{e \in \Omega} p(e)e$$

With such a closed form, the computations involved in dealing with random variables are substitution and differentiation, both of which can be carried out with a symbolic mathematics environment.

4.2 Computing Closed Forms for Series

Results from formal series theory (see, for example, [4]) tell us that if Ω is a rational set, and if $p(e)$ is defined as the product of the probabilities of the elementary results in A, then the series:

$$E = \sum_{e \in \Omega} p(e)e$$

is rational, meaning that it can be expressed as the quotient of two polynomials in the variables $a \in A$.

One way to compute this quotient is first to relabel the automaton recognizing Ω by multiplying each result r by its probability $p(r)$. For example, the automaton of Fig. 1 is relabeled in Fig. 5.

We can represent this new automaton with a matrix M indexed by its states, and whose entry (i, j) is the formal sum of the expressions $p(r)r$ for each transition between states i and state j.

For example, the matrix corresponding to the automaton of Fig. 5 is the following:

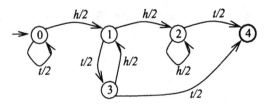

Fig. 5. Relabeling an automaton

$$M = \begin{bmatrix} t/2 & h/2 & 0 & 0 & 0 \\ 0 & 0 & h/2 & t/2 & 0 \\ 0 & 0 & h/2 & 0 & t/2 \\ 0 & h/2 & 0 & 0 & t/2 \\ 0 & 0 & 0 & 0 & 0 \end{bmatrix}$$

If we multiply the matrix M by itself k times, then the (i, j) entry of M^k will be the formal sum of all terms obtained by taking the product of labels of all paths of length k from state i to state j. For example, the entry $(0, 2)$ of M^3 would be:

$$\frac{thh}{8} + \frac{hhh}{8}$$

Let M^* be the infinite sum $I + M + M^2 + M^3 + \dots$. Then the entry (i, j) of M^* contains the formal sum:

$$\sum p(e)e$$

where the sum is taken over all paths from state i to state j. Thus the entry $(0, n)$ of M^*, where 0 is the initial state and n is the final state, is exactly:

$$E = \sum_{e \in \Omega} p(e)e$$

The matrix M^* is also given by $(I - M)^{-1}$ since

$$M^*(I - M) = M^* - MM^* = I$$

We can thus obtain the value of M^* by inverting the matrix $(I - M)$. This can be done with the help of a symbolic mathematics environment such as Maple. In the case of the coin tossing experiment we get:

$$E = M^*{}_{(0,4)} = \frac{1}{2-t}h\frac{1}{4-th}tt + \frac{1}{2-t}h\frac{1}{4-th}h\frac{2}{2-h}t$$

4.3   Random Variables and Other Computations

A *linear* random variable is the restriction to Ω of a function $V : A^* \to \mathbb{N}$ such that $V(ef) = V(e) + V(f)$. Examples of linear random variable are the length of a word, or the number of occurrences of a symbol in a word. If V is a random variable, then its generating function can be obtained by replacing each symbol $r \in A$ by $x^{V(r)}$ in the series representing the experiment. For example, the generating series of the *length* in the coin tossing example is given by:

$$F(x) = \frac{x^3}{(2-x)(4-x^2)} + \frac{x^3}{(2-x)(4-x^2)(2-x)}$$

thus, the mean length of the experiment is equal to $F'(x)_{x=1} = 5\frac{1}{3}$.

The probability of events that can be described by automata can also be computed using the same techniques. For example, let X be the event *"the pattern hht shows up first"*. Then X can be described by the intersection of the sets recognized by the automata of Fig. 1 and Fig. 3. Since this intersection is again rational, we can evaluate:

$$\sum_{e \in X} p(e)e = \frac{1}{2-t}h\frac{1}{4-th}h\frac{2}{2-h}t$$

and compute the probability of X by replacing h and t by 1 in this expression, yielding $\frac{2}{3}$.

5   Technical Aspects

The core of the laboratory is written in JAVA. It handles interactions with the user such as the definition of trials and experiments, and occasionaly calling Maple to solve symbolic matrix inversion and differentiation. In order to set up an experiment, one must first enumerate the results of trial, and their associated probabilities:

```
CoinFlip <- DefineTrial({h, t}; p(h) = 1/2, p(t) = 1/2).
```

An experiment is defined by a trial and a stopping condition, such as:

```
Experiment1 <- DefineExperiment(CoinFlip, htt + hht).
```

Answering questions about an experiment often involves the definition of a random variable. For example, the length can be defined by the call:

```
length <- DefineVariable(Experiment1, freq(h) + freq(t))
```

and one can obtain either its expectation, or probabilities related to this variable such as:

```
Probability(length = 4)?
```

The laboratory can also compute the probability of an event defined with the expressions used for stopping conditions. For example, the probability that the experiment stops with *hht* is computed by the call:

    Probability(hht)?

Small experiments are handled efficiently. The main computational problem is matrix inversion. The size of matrices is equal to the number of states of automata describing the possible outcomes of an experiment. The first and third constructions of Section 3.2 yield *reasonably* small automata: in the first case, the number of states is given by the length of the word s, and in the third case, it is given by the sum of the number of states of A_1 and A_2. The second construction, associating an automaton to $E_1 + E_2$, involves a product that can be responsible of a polynomial growth of the number of states.

The current implementation of the laboratory is quite straightforward. However, while trying to find the computational limits of this implementation, we had to construct far stretched experiments.

For example, consider the coin tossing experiment in which the stopping condition is to obtain 60 consecutive *tails*:

$$tttttttttttttttttttttttttttttttttttttttttttttttttttttttttttt.$$

In 5 minutes, the laboratory gave the expected length of the experiment as:

$$2305843009213693950$$

Table 1 gives a compilation of the expected length and time of computation for the experiments in which the stopping condition is to obtain k consecutive *tails*.

Table 1. Expected length and computation time for k consecutive *tails*

k	Expected length	Time (sec)
2	6	6
10	2046	7
20	2097150	10
30	2147483646	22
40	2199023255550	46
50	2251799813685246	104
60	2305843009213693950	274
70	2361183241434822606846	489
80	2417851639229258349412350	1003

Since the stopping condition for these experiments yields a very simple matrix, we suspected that more complex patterns would tax the resources of Maple. Indeed, in computing the expected length of experiments in which the stopping condition is to obtain a sequence of alternating h and t of length k, we had to stop at $k = 50$. The results for smaller values are given in Table 2.

Table 2. Expected length and computation time for k alternating h and t

k	Expected length	Time (sec)
2	4	6
10	1364	7
20	1398100	16
30	1431655764	69
40	1466015503700	331

6   Conclusion

Our initial goal was to develop tools for computational probabilities. Using semaphore codes as a suitable class of languages to describe the possible outcomes of an experiment, we were able to construct a computer environment in which random experiments can be defined, and queries about probabilities of events are automatically answered.

The current implementation is not optimized. Nevertheless, the scope of experiments that can be solved is impressive, thanks to Maple. Further developments include the availability of the laboratory on the Internet, and the exploration of other classes of languages that could yield interesting computational results.

References

1. Aho, A., Hopcroft, J., Ullman, J., The Design and Analysis of Computer Algorithms, Addison-Wesley, 1974.
2. Beaulieu, M.C., Langages rationnels et probabilités discrètes, Mémoire de Maitrise, Université du Québec à Montréal, 1998.
3. Berstel, J., Perrin, D., Theory of Codes, Academic Press, 1985.
4. Berstel, J., Reutenauer, C., Les séries rationnelles et leurs langages, Masson, 1984.
5. Eilenberg, S., Automata, Languages and Machines, Vol A, Academic Press, 1974.
6. Flajolet, P., Salvy, B., Zimmermann, P., *Automatic Average-Case Analysis of Algorithms*, Theoretical Computer Science 79, 1991, 37-109.
7. Geniet, D., Thimonnier L., *Using Generating Functions to Compute Concurrency*, Fundamental of Computation Theory, LNCS 380, 1989, 185-196.
8. Hansel, G., Perrin, D., *Rational Probability Measures*, Theoretical Computer Science, 65, 1989, 171-188.
9. Penney, W., *Problem 95: Penney-Ante*, Journal of Recreational Mathematics, 7, 1974, 321.
10. Schtuzenberger, M. P., *On the synchronizing properties of certain prefix codes*, Inform. and Control 7, 1964, 23-36.
11. Shannon, C., The Mathematical Theory of Communication, The University of Illinois Press, Urbana, 1949.

Automata and Binary Decision Diagrams

Jean-Francis Michon and Jean-Marc Champarnaud

LIFAR
Université de Rouen, Faculté des Sciences et des Techniques
76821 Mont-Saint-Aignan Cedex, France
{michon, champarnaud}@dir.univ-rouen.fr

Abstract. We show that the concept of automata minimization leads to
a nice interpretation of the famous canonicity of binary decision diagrams
discovered by Bryant.

1   Introduction

The aim of this paper is to enlighten the links between automaton minimization and the construction of binary decision diagrams of Boolean functions. We
give some direct applications to complexity. Section 2 recalls basic definitions
regarding the theories of Boolean functions and automata. Section 3 deals with
complexity results. Section 4 asks some questions.

2   Definitions and Conventions

2.1   Boolean Functions

In this field it is convenient to introduce mathematical structure "just in time"
so we choose this poor definition:

Definition 1 *A Boolean function on an arbitrary set A is a map from A to the
set of integers $B = \{0, 1\}$.*

So a Boolean function can be identified with its graph which is a part of $A \times \{0, 1\}$.
This graph is a set of pairs, called the truth table of the Boolean function. In
computer science we don't deal with arbitrary sets A but with sets of n-tuples
of bits. The set A is $B^n = \{0, 1\}^n$ for some integer $n > 0$. We shall assume that
we are dealing with those sets only. We would like to stress that this is not a
light assumption: it means that the set A is very special set with plenty of magic
properties.

This implies that a Boolean function is now a function of n variables each
ranging over B and with values in B:

$$(x_1, ..., x_n) \in B^n \mapsto f(x_1, ..., x_n) \in B$$

We have a canonical order on B induced by that of $N : 0 < 1$. This order on
B induces lexicographically a total order on A. This permits one to present the

J. M. Champarnaud, D. Maurel, D. Ziadi (Eds.): WIA '98, LNCS 1660, pp. 178–182, 1999.

truth table of any Boolean fonction f as the well known ordered table with 2^n rows and 2 columns. The classical other way to present the truth table is that of a binary ordered tree. Its construction is as follows:

a) The root vertex is labeled with x_1.

b) The two sons of any vertex labeled x_i are labeled x_{i+1} if $i < n$.

c) The two sons of the vertices labeled x_n are leaves.

d) The left (resp. right) edge starting from an interior vertex is labeled with 0 (resp. 1).

e) Each of the 2^n leaves is labeled with the value of f on the n-tuple corresponding to the unique path from the root x_1 to this leaf.

We call this tree the canonical tree associated with f and denote it $T(f)$.

2.2 Binary Decision Diagrams

A binary decision diagram is a member of a family of graphs containing binary trees like $T(f)$ and all graphs inductively reduced by the following two rules:

(Reduction 1) Identify two isomorphic subtrees G and H by deleting one and redirecting incoming edges to the root of the other.

(Reduction 2) If a vertex v has exactly one son, suppress v and the two edges starting from it and connect its incoming edges The graphs obtained at each step are always directed, acyclic, single originated; their branching index is at most two and they have one or two terminal vertices respectively labeled 0 and 1. This process is confluent and terminates. Applied to $T(f)$ the result is called the BDD or, more precisely, the ROBDD (reduced ordered BDD) of f. The complexity of a BDD is defined as the number of its vertices. Sometimes the complexity is very low and sometimes exponential in the number of variables. When the complexity of a BDD is low computers can work with and manipulate the Boolean function it refers to. This subject is of great interest in verification and also in reliability theory. We refer the reader to [BR, BR2] for excellent accounts and detailed explanations.

Definition 2 *The quasi-reduced BDD of a Boolean function is the graph obtained by iterating reduction steps of type 1 until this reduction cannot be applied any longer.*

We shall speak of the QRBDD of f.

2.3 Automata

Let X be a finite alphabet. An *automaton* over X is a 4-tuple $\mathcal{A} = (Q, I, F, E)$ where Q is a set of *states*, I is a subset of Q whose elements are the *initial states*, F is a subset of Q whose elements are the *final states*, (or *terminal states*), E is a subset of the cartesian product $Q \times X \times Q$ whose elements are the *edges*. Let $\mathcal{A} = (Q, I, F, E)$ be an automaton. A *path* of $\mathcal{A}$ is a sequence (q_i, a_i, q_{i+1}), $i = 1, \ldots, n$, for some $n \geq 1$, of consecutive edges. Its *label* is the word $w = a_1 a_2 \ldots a_n$. A word $w = a_1 a_2 \ldots a_n$ is *recognized* by the automaton $\mathcal{A}$ if there is a path with label w such that $q_1 \in I$ and $q_{n+1} \in F$. The language *recognized* by the automaton $\mathcal{A}$ is the set of words which it recognizes.

We now add some extra structure to $T(f)$ to equip it with the structure of an automaton.

Definition 3 *The truth automaton $\mathcal{T}(f) = (Q, I, F, E)$ of f over the alphabet $\{0, 1\}$ deduces from $T(f)$ as follows:*
1) Q is made of all the vertices of $T(f)$.
2) The unique initial state is the root of $T(f)$.
3) F is the set of all the leaves of $T(f)$ labeled 1.
4) The edges in E are the edges of $T(f)$, with the same labels.

The truth automaton of f can be viewed as the canonical disjunctive normal form of f. This automaton exactly recognizes the n-tuples of B^n where f takes the value 1 (called the support of f). In other terms the support of f is the (regular) language recognized by the automaton $\mathcal{T}(f)$. This automaton is obviously deterministic and accessible, but it is not complete. The completion of $\mathcal{T}(f)$ can be canonically performed by adding a non-final extra state (a sink) and connecting every leaf (and this sink) to the sink by each of the letters 0 and 1. We then loose the tree structure of the underlying graph. We are now in the position to minimize this automaton; we shall denote the minimal automaton by $\mathcal{T}'(f)$.

3   Minimal Automata and BDDs

Theorem 1. *The quasi reduced BDD of f is isomorphic to $\mathcal{T}'(f)$.*

Proof. This is a consequence of the Nerode equivalence [NE], which can be stated as follows: two vertices are said to be Nerode equivalent if the sub-automata rooted in these two vertices recognize the same language.
Suppose two vertices of $\mathcal{T}(f)$ are Nerode equivalent. This is exactly equivalent with the fact that the subtrees rooted at each of these vertices are isomorphic.

The reduction step of type 2 of ROBDDs is proper to Boolean algebra: it means that if the substitution of x_i by 0 or 1 does not affect the value of a Boolean function then the function is independent of x_i, the variable is "dumb". This is not exactly the same for non Boolean functions. So it's not surprising that the reduction of type 2 cannot be captured by the theory of minimization. The truth automaton $\mathcal{T}(f)$ of f being acyclic, the minimal automaton $\mathcal{T}'(f)$ can be computed in time linear in terms of the number of states of $\mathcal{T}(f)$ [RE]. So the complexity of computing the QRBDD of f starting from $\mathcal{T}(f)$ is $O(2^n)$. Of course, this is quite unrealistic because a Boolean function is never given by its truth table when the number of variables is large.

We now consider the complexity of ROBDDs and QRBDDs. Akers [AK] says that the worst case ROBDDs complexity is $O(2^n/n)$. The result is made more precise in [LL] where this worst case complexity is proved to be less than $(2^n/n)(2+\varepsilon)$ as n grows and ε is any arbitrarily small strictly positive real number. Some further results are given in [W].

An interesting result, coming from automaton theory, is given by Champarnaud and Pin [CP]. Translated into our context, it says that the maximal number of states, denoted by $g(n)$, of any QRBDD is exactly

$$g(n) = \sum_{1 \leq i \leq n} \min(2^i, 2^{2^{n-i}} - 1)$$

and

$$\liminf_{n \to \infty} ng(n)/2^n = 1,$$

$$\limsup_{n \to \infty} ng(n)/2^n = 2.$$

For a given n, an explicit construction of all the QRBDDs with a maximal number of states is also suggested. The construction implies the result of [LL] for ROBDDs because reduction steps of type 2 can occur only in the tail part of their automaton (where the number of states decays doubly exponentially). The reduction of type 2 is, consequently, negligible.

A side consequence is that we can explicitly give a family of very complex Boolean functions from the point of view of BDDs. Boolean functions under permutations of variables.

4   Conclusion

The conclusion is that automata are one essential part of BDD theory and that this part is quite independent of Boolean algebra. It follows that the same process (minimization) can be applied to more general BDD-like structures for integer valued functions, arithmetic functions ... Conversely the BDD theory suggests good ideas to automata specialists. For example, we don't touch here the crucial problem of BDDs which is the sensitivity of its complexity to the ordering of variables [FS]. This must be reflected in automata theory by the notion of commutative equivalence of languages. In the same type of idea the process of identifying isomorphic subtrees can be viewed as a Fourier-Hadamard transformation (called the spectrum of a Boolean function). This suggests there may exist a sort of "Fourier transform" under the minimization process for automata.

References

[AK] S. B. Akers, Binary Decision Diagrams, *IEEE Transactions on Computers* **C-27**(6) (1978) 509–516.

[BR] R. E. Bryant, Graph Based Algorithms for Boolean Functions Manipulation, *IEEE Transactions on Computers* **C-35**(8) (1986) 677–691.

[BR2] R. E. Bryant, Binary Decision Diagrams and beyond, *Survey Report*, Carnegie Mellon University.

[CP] J.-M. Champarnaud and J.-E. Pin, A Maxmin Problem on Finite Automata, *Discrete Applied Mathematics* **23** (1989) 91–96.

[FS] S. J. Friedman, K. J. Supowit, Finding the Optimal Ordering for Binary Decision Diagram, *IEEE Transactions on Computers* **39**(5) (1990).

[LL] H.-T. Liaw and C.-S. Lin, On the OBDD Representation of General Boolean Functions, *IEEE Transactions on Computers* **C-41**(46) (1992) 661–664.

[NE] A. Nerode, Linear Automata Transformation, Proc. *AMS 9* (1958) 541–544.

[RE] D. Revuz, Minimization of Acyclic Deterministic Automata in Linear Time, *Theoret. Comput. Sci.* **92** (1992) 181–189.

[W]   I. Wegener, Efficient Data Structure for Boolean Functions, *Discrete Mathematics* **136** (1994) 347–372.

Operations over Automata with Multiplicities

Gérard Duchamp, Marianne Flouret, and Éric Laugerotte

LIFAR
Université de Rouen, Faculté des Sciences et des Techniques
76821 Mont-Saint-Aignan Cedex, France
{Gerard.Duchamp, Marianne.Flouret, Eric.Laugerotte}@dir.univ-rouen.fr

Abstract. We present here theoretical results coming from the implementation of the package called AMULT (automata with multiplicities in
several noncommutative variables). We show that classical formulas are
"almost every time" optimal and characterize the dual laws preserving
rationality.

1 Introduction

Noncommutative formal series (i.e. functions on the free monoid, with values in a
- commutative or not - semiring) encode an infinity of data. Rational series can be
represented by linear recurrences, corresponding to automata with multiplicities,
and therefore they can be generated by finite state processes. Literature can
be found on these "weighted automata" (e.g. [10], [13]) and their theoretical
(e.g. [8]) and practical (e.g. [2], [12]) applications (recently one of us solved a
conjecture in operator theory using these tools [3]). The theory was founded by
Schützenberger in 1961 [14] where the link between recognizable and rational
series is showed (see also [15]), extending to rings (and to semirings [1]) Kleene's
result for languages [9] (corresponding to boolean coefficients). In 1974, for the
case of fields, Fliess [5] extended the proof of the equivalence of minimal linear
representations, using Hankel matrices. All these results allow us to construct
an algorithmic processing for this series and their associated operations. In fact,
classical constructions of language theory have multiplicity analogues which can
be used in every domain where linear recurrences between words are handled.

All these operations can be found in the package over automata with multi-
plicities (called AMULT). This package is a component of the environment SEA
(Symbolic Environment for Automata) under development at the University of
Rouen.

The paper here is devoted to the study of polynomial compositions of linear
recurrences. We point out two facts directly linked to implementation. The first
result says that the classical formulas are "almost everywhere" optimal (this is
clear from tests at random). The second point shows that three laws known to
preserve rationality are of the same nature: they arise by dualizing alphabetic
morphisms. Moreover, they are, up to a deformation, the only ones of this kind,
which of course, shows immediately in the implemented formulas.

J.-M. Champarnaud, D. Maurel, D. Ziadi (Eds.): WIA'98, LNCS 1660, pp. 183–191, 1999.

2   Preamble

Let $K\langle\langle A\rangle\rangle$ be the set of noncommutative formal series with A a finite alphabet and K a semiring. A series denoted $S = \sum_{w\in A^*}\langle S|w\rangle w$ is recognizable iff there exists a row vector $\lambda \in K^{1\times n}$, a morphism of monoids $\mu : A^* \to K^{n\times n}$ and a column vector $\gamma \in K^{n\times 1}$, such that for all $w \in A^*$, one has $\langle S|w\rangle = \lambda\mu(w)\gamma$. Throughout the paper, we will denote by $S : (\lambda, \mu, \gamma)$ this property and say that (λ, μ, γ) is a linear representation of S. The integer n is called the *rank* of the linear representation (λ, μ, γ) [5].

Let $K^{\mathrm{rat}}\langle\langle A\rangle\rangle$ be the set of rational noncommutative formal series, that is the set generated from the letters and the laws "." (concatenation or Cauchy product), $*$ (star operation, partially defined), $\times$ (external product) and $+$ (union or sum). The following important theorem for series [14] is the analogue of Kleene's theorem for languages.

Theorem 1 (Schützenberger, 1961). *A formal series is recognizable if and only if it is rational.*

A reduced linear representation $S : (\lambda, \mu, \gamma)$ is a linear representation of minimal dimension among all its representations. Existence is assumed by definition, unicity is proved in case K is $I\!B$ or a (commutative or not) field [7] but is problematic in general. This minimum is called the *rank* of the series S [14]. In case K is a field, the rank of S is the dimension of the linear span of the shifts of S (see Sect. 3). It is the smallest number of nodes of an automaton with behaviour S. Here, minimization (up to an equivalence) is possible [14] (see also [1]). An explicit algorithm is given in full details in [7] (notice that this algorithm is valid as well for noncommutative multiplicities) as well as the construction of intertwining matrices.

Notice that the specialisation of K to the boolean semiring $I\!B$ yields to the case of classical finite state automata.

3   Constructing Usual Laws

3.1   Operations on Linear Representations

We expound here universal formulas for constructing linear representations. They can be applied to any semiring K. For two representations of ranks n and m, it will be provided a representation of rank $r(n, m)$. Let us recall some classical facts. Classical operations on series are sum, external product and star (unary and partially defined). By definition, the sum of two series R and S is

$$R + S = \sum_{w\in A^*} \left(\langle R|w\rangle + \langle S|w\rangle\right)w \ ,$$

their concatenation (or Cauchy product)

$$R.S = \sum_{w\in A^*} \left(\sum_{uv=w} \langle R|u\rangle\langle S|v\rangle\right) w \ ,$$

and the star of a series S

$$S^* = \sum_{n \geq 0} S^n = 1 + SS^*$$

if its constant term is zero (such a series is said to be proper). The preceding operations have polynomial counterparts in terms of linear representations. We gather them in the following proposition.

Proposition 1 Let $R : \rho_r = (\lambda^r, \mu^r, \gamma^r)$ (resp. $S : \rho_s = (\lambda^s, \mu^s, \gamma^s)$) of rank n (resp. m). The linear representations of the sum, the concatenation and the star are respectively
$R + S$:

$$\rho_r \boxplus \rho_s = \left(\left(\lambda^r \ \lambda^s \right), \left(\begin{array}{c|c} \mu^r(a) & 0 \\ \hline 0 & \mu^s(a) \end{array} \right)_{a \in A}, \left(\begin{array}{c} \gamma^r \\ \gamma^s \end{array} \right) \right) , \tag{1}$$

$R.S$:

$$\rho_r \boxdot \rho_s = \left(\left(\lambda^r \ 0 \right), \left(\begin{array}{c|c} \mu^r(a) & \gamma^r \lambda^s \mu^s(a) \\ \hline 0 & \mu^s(a) \end{array} \right)_{a \in A}, \left(\begin{array}{c} \gamma^r \lambda^s \gamma^s \\ \gamma^s \end{array} \right) \right) , \tag{2}$$

If $\lambda^s \gamma^s = 0$, S^* :

$$\rho_s^{\boxtimes} = \left(\left(0 \ 1 \right), \left(\begin{array}{c|c} \mu^s(a) + \gamma^s \lambda^s \mu^s(a) & 0 \\ \hline \lambda^s \mu^s(a) & 0 \end{array} \right)_{a \in A}, \left(\begin{array}{c} \gamma^s \\ 1 \end{array} \right) \right) . \tag{3}$$

Remark 1. 1. Formulas (1) and (2) provide associative laws on triplets. They can be found explicitly in [2].

2. Formula (3) makes sense even when $\lambda^s \gamma^s \neq 0$ (this fact will be used in the density result of the Sect. 3.2).

3.2   Sharpness

Here we discuss the sharpness of the preceding constructions. Indeed, testing our package showed us that "almost everytime" the compound linear representation was minimal when the data were choosen at random. The crucial point in the proof of Theorem 2 is the fact that certain polynomial indicators are not trivial. For this, we use suited examples which are gathered in the following subsection.

Test Automata Let $\mathcal{B} = (S_i)_{1 \leq i \leq n}$ be a finite sequence of series generating a stable module and $S = \sum_{i=1}^n \lambda_i S_i$. It is well known that the triplet

$$\left(\sum_{i=1}^n \lambda_i e_i, \ \left([\mu_{i,j}(a)]_{1 \leq i,j \leq n} \right)_{a \in A}, \ \sum_{i=1}^n \langle S_i | 1 \rangle e_i^* \right)$$

(where $e_i := (0, \cdots, 1, \cdots 0)$ with the entry 1 at place i, e_i^* the transpose of e_i, and $a^{-1}S_i = \sum_{j=1}^{n} (\mu(a))_{ij} S_j$ for any letter $a \in A$) is a linear representation of S.

Here, to each series of one variable, $S = \sum_{p \geq 0} \alpha_p a^p$, of rank n, over a field K, we associate the triplet $\tau(S)$ given by $\mathcal{B} = (a^{-p} S)_{0 \leq p \leq n-1}$, of course $S \in K\langle\langle A \rangle\rangle$ providing that a belongs to A. This will not affect the rank.

Lemma 1 Let $S_{\alpha,n} = \dfrac{1}{(1 - \alpha a)^n}$ and $T_n = \dfrac{a^{n-1}}{1 - a^n}$ be $\mathbb{Q}$-series.

1. The rank of $S_{\alpha,n}$, $S_{\alpha,n} + S_{\beta,m}$ $(\alpha \neq \beta)$, and $S_{\alpha,n}.S_{\alpha,m}$ are respectively n, $n + m$ and $n + m$.
2. The rank of T_n is n and that of T_n^* is $n + 1$.

Proof. Straightforward. □

Density The following theorem proves that if the data are choosen "at random" in bounded domains, the compound automaton is almost surely minimal. More precisely :

Theorem 2. Let A finite and $\rho_1 = (\lambda_1, \mu_1, \gamma_1), \rho_2 = (\lambda_2, \mu_2, \gamma_2)$ two linear representations, be choosen "at random" within bounded non trivial disks of K ($\mathbb{R}$ or $\mathbb{C}$). Then the probability that the linear representation $\rho_1 \boxplus \rho_2$ (resp. $\rho_1 \boxdot \rho_2$, $\rho_1 ^{\boxtimes}$) be minimal is 1.

Proof. The proof rests on the following lemma.

Lemma 2 There is a polynomial mapping $P : K^{|A| \times n^2 + 2n} \to K^s$ such that $P(\lambda, \mu, \gamma) = 0$ iff (λ, μ, γ) (a linear representation of degree n) is not minimal.

Proof. By a theorem of Schützenberger [14], the representation (λ, μ, γ) is minimal iff $\lambda\mu(K\langle A \rangle) = K^{1 \times n}$ (resp. $\mu(K\langle A \rangle)\gamma = K^{n \times 1}$), as there is a prefix (resp. suffix) part $U \subset A^*$ (resp. $V \subset A^*$) such that $\lambda\mu(U)$ (resp. $\mu(V)\gamma$) is a basis, we have $U \subset A^{<n}$ (resp. $V \subset A^{<n}$). Let $A^{<n} = \{w_1, w_2, \cdots, w_m\}$ (denoting $w_1 := 1$), one constructs the $m \times n$ (resp. $n \times m$) matrix $L = \begin{pmatrix} \lambda\mu(w_1) \\ \lambda\mu(w_2) \\ \vdots \\ \lambda\mu(w_m) \end{pmatrix}$

(resp. $M = (\mu(w_1)\gamma \cdots \mu(w_m)\gamma))$, these matrices have polynomial entries in the data. In view of what precedes, minimality is equivalent to the non nullity of some $n \times n$-minor of L and of M. Sorting these minors as a vector, one get the desired polynomial mapping $K^{|A| \times n^2 + 2n} \to K^s$ with $s = 2 \binom{m}{n}$. □

The other steps go as follows.

1. For the two first operations, let $P_{\boxplus} = P(\rho_1 \boxplus \rho_2)$, $P_{\boxdot} = P(\rho_1 \boxdot \rho_2)$, and prove that $P_{\boxplus}$ (resp. $P_{\boxdot}$) is not trivial using $\tau(S_{\alpha,n}) = \rho_1$ and $\tau(S_{\beta,n}) = \rho_2$, $\alpha \neq \beta$ (resp. $\tau(S_{\alpha,n}) = \rho_1$ and $\tau(S_{\alpha,m}) = \rho_2$). For the star operation, prove that $P_{\boxtimes} = P(\rho_1^{\boxtimes})$ is not trivial using $\tau(T_n) = \rho_1$.

2. End of the proof: if $\phi : K^r \to K^s$ is polynomial and not trivial, let ν be the normalized uniform probability mesure on the product of disks, then the probability such that $\phi(\nu) \neq 0$ is 1 as $\phi^{-1}\{0\}$ is closed with empty interior.

□

4   Dual Laws

Let $a, b \in A$, $u, v \in A^*$, and $\odot_{\epsilon,q}$ be the law defined recursively by

$$\begin{cases} 1 \odot_{\epsilon,q} 1 = 1, \ a \odot_{\epsilon,q} 1 = 1 \odot_{\epsilon,q} a = a, \\ au \odot_{\epsilon,q} bv = \epsilon\big(a(u \odot_{\epsilon,q} bv) + b(au \odot_{\epsilon,q} v)\big) + q\delta_{a,b}a(u \odot_{\epsilon,q} v) \end{cases}$$

with $\delta_{a,b}$ the Kronecker delta.

One immediately checks that this law is associative iff $\epsilon \in \{0,1\}$. We get the well-known shuffle ($\sqcup\!\sqcup = \odot_{1,0}$), infiltration ($\uparrow = \odot_{1,1}$) and Hadamard ($\odot = \odot_{0,1}$) products ([4], [11]). Then, $\odot_{1,q}$ is a continuous deformation between shuffle and infiltration. These laws can be called "dual laws" as they proceed from the same template that we now describe. We use an implementable realisation of the lexicographically ordered tensor product. Let us recall that the tensor product of two spaces U and V with bases $(u_i)_{i\in I}$ and $(v_j)_{j\in J}$ is $U \otimes V$, with the basis $(u_i \otimes v_j)_{(i,j)\in I\times J}$. For the sake of computation, we impose that the set $I \times J$ be lexicographically ordered.

Let $K\langle A\rangle \otimes K\langle A\rangle$ be the "double" non commutative polynomial algebra that is the set of finite sums $P = \sum_{u,v\in A^*}\langle P|u\otimes v\rangle u\otimes v$, the product being given by $(u_1 \otimes v_1)(u_2 \otimes v_2) = u_1 u_2 \otimes v_1 v_2$. The construction of dual laws is based on the following pattern.

Let $c : K\langle A\rangle \to K\langle A\rangle \otimes K\langle A\rangle$, if for all $w \in A^*$, the set $\{w : \langle u\otimes v|c(w)\rangle \neq 0\}$ is finite (in which case c will be called *locally finite*), then the sum

$$u \ \square_\alpha \ v = \sum_{w\in A^*} \langle u \otimes v|c_\alpha(w)\rangle w$$

exists and defines a (binary) law $\square_\alpha$ on $K\langle A\rangle$, dual to c_α. This extends to series by

$$\langle R \ \square_\alpha \ S|w\rangle := \langle R \otimes S|c_\alpha(w)\rangle .$$

One can show easily that the three laws $\odot$, $\sqcup\!\sqcup$ and $\uparrow$ come from coproducts defined on the words by

1. $c_\alpha(a_1 a_2 \cdots a_n) = c_\alpha(a_1)c_\alpha(a_2)\cdots c_\alpha(a_n)$,
2. $c_\odot(a) = a \otimes a$, $c_{\sqcup\!\sqcup}(a) = a \otimes 1 + 1 \otimes a$, $c_\uparrow(a) = a \otimes 1 + 1 \otimes a + a \otimes a$,

and, generally, $c_{\epsilon,q}(a) = \epsilon(a\otimes 1 + 1\otimes a) + qa\otimes a$. Moreover the coproducts $c_{\odot_{1,q}}$, are the only alphabetic morphisms for which the laws are associative with 1 as unit. More precisely:

Proposition 2 *Let K be a field, and $c_\alpha : K\langle A \rangle \to K\langle A \rangle \otimes K\langle A \rangle$ the alphabetic morphism defined on the letters of A by*

$$c_\alpha(a) = \sum_{p,q \geq 0} \alpha_{p,q} a^p \otimes a^q$$

with $c_\alpha(1) = 1 \otimes 1$.

1. *The morphism c_α is locally finite iff $\alpha_{0,0} = 0$.*
2. *Providing $\alpha_{0,0} = 0$, the following assertions are equivalent.*
   (a) *The law $\square_\alpha$ defined by $\langle u \square_\alpha v | w \rangle := \langle u \otimes v | c_\alpha(w) \rangle$ $(u, v, w \in A^*)$ is associative.*
   (b) *The coefficients $\alpha_{p,q}$ satisfy the relations $\alpha_{p,q} = 0$ for p or $q \geq 2$, $\alpha_{0,1}, \alpha_{1,0} \in \{0,1\}$ and $\alpha_{0,1}\alpha_{1,1} = \alpha_{1,0}\alpha_{1,1}$.*
3. *Providing (2b), the element 1_{A^*} is a unit for $\square_\alpha$ iff $\alpha_{0,1} = \alpha_{1,0} = 1$.*

Proof. 1. We have $c_\alpha(a) = \alpha_{0,0} 1 \otimes 1 + \sum_{p+q \geq 1} \alpha_{p,q} a^p \otimes a^q$, and then for all $n \geq 0$,

$$c_\alpha(a^n) = \alpha_{0,0}^n 1 \otimes 1 + \sum_{p+q \geq 1} \beta_{p,q} a^p \otimes a^q \text{ for some } \beta_{p,q}.$$ If $\alpha_{0,0}$ were not zero, the term $1 \otimes 1$ would appear in an infinity of words, and then c_α would not be locally finite.

Conversely, if $\alpha_{0,0} = 0$, then $c_\alpha(a) = \sum_{p+q \geq 1} \alpha_{p,q} a^p \otimes a^q$ and for all word $w = a_1 \cdots a_n \in A^*$,

$$c_\alpha(w) = \sum_{\substack{p_i + q_i \geq 1 \\ 1 \leq i \leq n}} \left(\prod_{i=1}^n \alpha_{p_i, q_i} \right) a_1^{p_1} \cdots a_n^{p_n} \otimes a_1^{q_1} \cdots a_n^{q_n}.$$

As $p_i + q_i \geq 1$, we have $\sum_{i=1}^n (p_i + q_i) \geq n$, that is to say $|w| \leq |u| + |v|$ and $Alph(w) = Alph(u) \cup Alph(v)$ with $u := a_1^{p_1} \cdots a_n^{p_n}$ and $v := a_1^{q_1} \cdots a_n^{q_n}$. The alphabet $Alph(u) \cup Alph(v)$ being finite, the set $\{w / (u \otimes v | c_\alpha(w)) \neq 0\}$ is finite.

2. First remark that the two assertions are equivalent to the condition

$$(Id \otimes c_\alpha) \circ c_\alpha = (c_\alpha \otimes Id) \circ c_\alpha \qquad (4)$$

The law $\square_\alpha$ is associative iff for all words $u_1, u_2, u_3 \in A^*$, we have

$$(u_1 \square_\alpha u_2) \square_\alpha u_3 = u_1 \square_\alpha (u_2 \square_\alpha u_3)$$

that is to say that, for all $w \in A^*$,

$$\langle (u_1 \square_\alpha u_2) \square_\alpha u_3 | w \rangle = \langle u_1 \square_\alpha (u_2 \square_\alpha u_3) | w \rangle .$$

But one has
$$\langle(u_1\Box_\alpha u_2)\Box_\alpha u_3|w\rangle = \langle(u_1\Box_\alpha u_2)\otimes u_3|c_\alpha(w)\rangle$$
$$= \langle u_1\otimes u_2\otimes u_3|(c_\alpha\otimes Id)\circ c_\alpha(w)\rangle$$
and
$$\langle u_1\Box_\alpha(u_2\Box_\alpha u_3)|w\rangle = \langle u_1\otimes(u_2\Box_\alpha u_3)|c_\alpha(w)\rangle$$
$$= \langle u_1\otimes u_2\otimes u_3|(Id\otimes c_\alpha)\circ c_\alpha(w)\rangle.$$

As u_1, u_2, u_3, w are arbitrary, we get $(c_\alpha\otimes Id)\circ c_\alpha = (Id\otimes c_\alpha)\circ c_\alpha$. To show the equivalence between (2b) and (4), suppose first that (4) holds. We endow $\mathbb{N}^k$ with the lexicographic order (reading from left to right for instance) which is compatible with addition and will be denoted $\prec$ (here, $k=2,3$). Then, if it is not zero, $c_\alpha(a)$ can be written

$$\alpha_{\bar{p},\bar{q}}a^{\bar{p}}\otimes a^{\bar{q}} + \sum_{(p,q)\prec(\bar{p},\bar{q})} \alpha_{p,q}a^p\otimes a^q \ ,$$

$(\bar{p},\bar{q})$ being the highest couple of exponents in the support. Then,
$$(c_\alpha\otimes Id)\circ c_\alpha(a) = \alpha_{\bar{p},\bar{q}}c_\alpha(a^{\bar{p}})\otimes a^{\bar{q}} + \sum_{(p,q)\prec(\bar{p},\bar{q})} \alpha_{p,q}c_\alpha(a^p)\otimes a^q$$
$$= \alpha_{\bar{p},\bar{q}}^{\bar{p}+1}a^{(\bar{p})^2}\otimes a^{\overline{pq}}\otimes a^{\bar{q}} + \sum_{(p,q,r)\prec(\bar{p}^2,\overline{pq},\bar{q})} \beta_{p,q,r}a^p\otimes a^q\otimes a^r,$$
but
$$(Id\otimes c_\alpha)\circ c_\alpha(a) = \alpha_{\bar{p},\bar{q}}a^{\bar{p}}\otimes c_\alpha(a^{\bar{q}}) + \sum_{(p,q)\prec(\bar{p},\bar{q})} \alpha_{p,q}a^p\otimes c_\alpha(a^q)$$
$$= \alpha_{\bar{p},\bar{q}}^{\bar{q}+1}a^{\bar{p}}\otimes a^{\overline{pq}}\otimes a^{(\bar{q})^2} + \sum_{(p,q,r)\prec(\bar{p},\overline{pq},\bar{q}^2)} \beta_{p,q,r}a^p\otimes a^q\otimes a^r.$$

Necessarily, $\bar{p}=\bar{p}^2$ and $\bar{q}=\bar{q}^2$, which is only possible when $\bar{p}\in\{0,1\}$ and $\bar{q}\in\{0,1\}$ and then $\alpha_{p,q}=0$ for p or $q\geq 2$. The equality now reads
$$\alpha_{1,0}a\otimes 1\otimes 1 + \alpha_{0,1}^2 1\otimes 1\otimes a + \alpha_{0,1}\alpha_{1,1}a\otimes 1\otimes a$$
$$=$$
$$\alpha_{1,0}^2 a\otimes 1\otimes 1 + \alpha_{0,1}1\otimes 1\otimes a + \alpha_{1,0}\alpha_{1,1}a\otimes 1\otimes a,$$
which implies (2b). The converse is a straightforward computation.

3. The condition 1_{A^*} is a unit for $\Box_\alpha$ implies that, for $a\in A$, we have
$$1\Box_\alpha a = a\Box_\alpha 1 = a \Leftrightarrow \langle 1\Box_\alpha a|a\rangle = \langle a\Box_\alpha 1|a\rangle = 1$$
$$\Leftrightarrow \langle 1\otimes a|c_\alpha(a)\rangle = \langle a\otimes 1|c_\alpha(a)\rangle = 1$$
$$\Leftrightarrow \begin{cases} \langle 1\otimes a|\sum_{p,q\geq 0}\alpha_{p,q}a^p\otimes a^q\rangle = 1 \\ \langle a\otimes 1|\sum_{p,q\geq 0}\alpha_{p,q}a^p\otimes a^q\rangle = 1 \end{cases}$$
$$\Leftrightarrow \alpha_{0,1} = \alpha_{1,0} = 1.$$
Now, this condition implies that, for each $w\in A^*$, $1\Box_\alpha w = w\Box_\alpha 1 = w$.

$\Box$

Remark 2. For just a commutative law the condition $\alpha_{0,1} = \alpha_{1,0}$ is sufficient. Moreover, the condition (2b) implies $\alpha_{0,1}, \alpha_{1,0}\in\{0,1\}$.

The preceding computation scheme has an immediate consequence on the implementation of the laws.

Proposition 3 *Let $R : (\lambda^r, \mu^r, \gamma^r)$ and $S : (\lambda^s, \mu^s, \gamma^s)$. Then*

$$R \,\square_\alpha\, S : (\lambda^r \otimes \lambda^s, \mu^r \otimes \mu^s \circ c_\alpha, \gamma^r \otimes \gamma^s) \ .$$

Proof. Computation.   □

4.1   Shuffle and Infiltration Product

Proposition 4 *Let $R : (\lambda^r, \mu^r, \gamma^r)$ (resp. $S : (\lambda^s, \mu^s, \gamma^s)$) with rank n (resp. m).*

1. *Representations of shuffle and infiltration products are respectively*

$$R \,{\sqcup\!\sqcup}\, S : (\lambda, \mu, \gamma) = (\lambda^r \otimes \lambda^s, (\mu^r(a) \otimes I_m + I_n \otimes \mu^s(a))_{a \in A}, \gamma^r \otimes \gamma^s) \ ,$$

   and

$$R \uparrow S : (\lambda^r \otimes \lambda^s, (\mu^r(a) \otimes I_m + I_n \otimes \mu^s(a) + \mu^r(a) \otimes \mu^s(a))_{a \in A}, \gamma^r \otimes \gamma^s) \ .$$

2. *The bound nm is sharp in both cases.*
3. *The density result of theorem 2 holds.*

Proof. Concerning point (2), an example reaching the bound for any rank is to consider the families of series $S_n = a^{n-1}$ and $T_n = b^{n-1}$ of rank n. The shuffle product $S_n {\sqcup\!\sqcup} S_m = a^{n-1} {\sqcup\!\sqcup} b^{m-1}$ ($a \ne b \in A$) has a minimal linear representation of rank nm. The same example is valid for the infiltration product as, for $a \ne b$, $a^n \uparrow b^m = a^n {\sqcup\!\sqcup} b^m$.   □

4.2   Hadamard Product

We recall that the Hadamard product ([6], [15]) of two series is the pointwise product of the corresponding functions (on words). We can use the machinery above to describe a representation of it.

Proposition 5 *Let $R : (\lambda^r, \mu^r, \gamma^r)$ (resp. $S : (\lambda^s, \mu^s, \gamma^s)$) with rank n (resp. m). A representation of the Hadamard product is*

$$R \odot S : \left(\lambda^r \otimes \lambda^s, (\mu^r(a) \otimes \mu^s(a))_{a \in A}, \gamma^r \otimes \gamma^s\right) \ ,$$

and the bound is asymptotically sharp.

Proof. Let $\beta(n, m) := \sup_{\substack{rank(R)=n \\ rank(S)=m}} rank(R \odot S)$. We claim that

$$\limsup_{n,m \to +\infty} \frac{\beta(n, m)}{nm} = 1 \ ,$$

(what we mean by "asymptotically sharp").
Indeed, let us consider the Hadamard product of two series of the family

$$S_n = \sum_{k \geq 0} a^{nk} = \frac{1}{(1 - a^n)} \ .$$

The rank of S_n is n, and

$$S_n \odot S_m = \sum_{k \geq 0} a^{nk} \odot \sum_{k' \geq 0} a^{mk'} = \sum_{p \geq 0} \langle S_n | a^p \rangle \langle S_m | a^p \rangle a^p$$
$$= \sum_{k \geq 0} a^{lcm(n,m)k} = S_{lcm(n,m)} \ .$$

Thus, for n and m coprime, the rank of the product is nm, which proves the claim.   □

References

1. Berstel, J., Reutenauer, C.: Rational Series and Their Languages. EATCS Monographs on Theoretical Computer Science, Springer-Verlag, Berlin (1988)
2. Culik II, K., Kari, J.: Finite state transformations of images. Proceedings of ICALP 95, Lecture Notes in Comput. Sci. **944** (1995) 51–62
3. Duchamp, G., Reutenauer, C.: Un critère de rationalité provenant de la géométrie non-commutative. Invent. Math. **128** (1997) 613–622
4. Eilenberg, S.: Automata, languages and machines, Vol. A, Acad. Press, New-York (1974)
5. Fliess, M.: Matrices de Hankel. J. Maths. Pures Appl. **53** (1974) 197–222
6. Fliess, M.: Sur divers produits de séries formelles. Bull. Sc. Math. **102** (1974) 181–191
7. Flouret, M., Laugerotte, E.: Noncommutative minimization algorithms. Inform. Process. Lett. **64** (1997) 123–126
8. Harju, T., Karhumäki, J.: The equivalence problem of multitape finite automata. Theoret. Comput. Sci. **78** (1991) 347–355
9. Kleene, S. C.: Representation of events in nerve nets and finite automata. Automata Studies, Princeton Univ. Press (1956) 3–42
10. Kuich, W., Salomaa, A.: Semirings, automata, languages. EATCS Monographs on Theoretical Computer Science, Springer-Verlag, (1986)
11. Lothaire, M.: Combinatorics on words. Addison-Wesley (1983)
12. Mohri, M., Pereira, F., Riley, M.: A rational design for a weighted finite-state transducer library. Proceedings of WIA'97, (1997) 43–53
13. Salomaa, A., Soittola, M.: Automata-theoretic aspects of formal power series. Springer-Verlag (1978)
14. Schützenberger, M. P.: On the definition of a family of automata. Inform. and Contr. **4** (1961) 245–270
15. Schützenberger, M. P.: On a theorem of R. Jungen. Proc. Amer. Soc. **13** (1962) 885–890

Paging Automata

Ricardo Ueda Karpischek

ueda@ime.usp.br
http://www.ime.usp.br/~ueda/biba/

Abstract. We present **biba**, a package designed to deal with representations of large automata. It offers a library able to build, even on a modest computer, automata where the sum of the numbers of states and edges achieves one billion or more. Two applications that use this library are provided as examples. They build the reduced automaton for a given vocabulary, and the suffix automaton of a given word. New programs can be developed using this library. In order to overcome physical memory limitations, **biba** implements a paging scheme, in such a way that the automata really reside on disk, making possible their permanent storage. Through a simple interface suited for **perl**, small scripts can be easily written to use and extract informations from these automata.

1 Introduction

Large and complex automaton-based data structures may arise when dealing with massive data sets. These automata can be viewed as ways to extract information from those sets, as well as ways to put and maintain them in a form suitable for specific applications, like search engines or spelling checkers. In this second case, the automaton is not a temporary need, but one definitive format under which the data set (dictionary) will be maintained.

Developed in a first moment to apply statistical tests in biological sequences [10], **biba** is mainly a **C** programming library, but it also contains programs built over this library offered as applications. It is able to make user-level *memory paging*, in the same sense that this term is used in operating systems textbooks (see for instance Bach [1]). However, **biba**'s paging implementation is not only a way to overcome physical memory limitations, but it is also and mainly an efficient way to permanently store automata on disk.

Paging will be a very interesting option in those cases where the automaton can compare to a large relational database, that resides on disk but is continuously being consulted and changed punctually along a large interval of time. An example we can think about is an automaton-based index for a large dynamic textual database, as **glimpse** [7] does.

In this paper we present the main features and characteristics of **biba**. A complete technical description may be found along the man pages and the source code, freely available in the URL given at the end of the paper. It was tested by the author in **x86** and **axp Linux**, but the library will compile and run with few or no changes in almost any 32 or 64-bit platform for which a **C** compiler is available.

J. M. Champarnaud, D. Maurel, and D. Ziadi (Eds.): WIA'98, LNCS 1660, pp. 192–198, 1999

2   An Overview from the User's Viewpoint

We introduce the features of **biba**, through some examples. For now, the visible part of **biba** are just two programs that integrate the package, **reduce** and **suffaut**.

The reduced automaton that recognizes a set of words, given one per line in a file, can be built using the **reduce** tool, that implements the algorithm due to Revuz [8]:

```
$ reduce words
```

The automaton just built resides in the pair of files **a.aut.st** (the states) and **a.aut.ed** (the edges). These are ordinary files, that you can copy, compress or transfer. Because of practical limitations that the user may need to deal with, we allow to split the files that contain an automaton in many parts, so one can combine the available areas of many filesystems, and take advantage of using disks in parallel. For instance, the following command will divide **a.aut.ed** in two parts, the first of them containing just 25 megabytes:

```
$ biba -s 25 a.aut.ed a.aut-1.ed
```

Next we present **suffaut**, a program that builds the suffix automaton of a given word, using the algorithm due to Blumer et alli [2]. In this same example we show how one paging parameter (among others) can be controlled through options. In this case we're specifying that the maximum ratio between total virtual memory size and total physical (RAM) memory size is 5. Just like in the first example, the automaton will reside in the files **a.aut.st** and **a.aut.ed**. We chose a binary word (the file /**bin**/**sh** seen as a sequence of bytes):

```
$ suffaut -p 5 -f /bin/sh
word length: 300668
states: 428275 (14 finals) edges: 636565
ram pages: 1300 virtual pages: 2141
```

This command took 64 seconds to complete in a 486 100Mhz with 32 megabytes of main memory. By default **suffaut** allocates 5 megabytes of RAM before allowing the ratio between virtual and physical memory becomes larger than 1. As the page size is 4096, **suffaut** allocated 1300 RAM pages. The final sizes for files **a.aut.st** and **a.aut.ed** were 5.5 and 3.2 megabytes, so **suffaut** allocated 13 bytes per state and 5 bytes per edge. These numbers will be explained in the next section (note that the algorithm requires two additional 32-bit fields per state).

Now we'll combine these two programs in order to test the accuracy of both. Let's consider *Good-de Bruijn* words over the alphabet $\{0,1\}$. The script **good** generates such a word with parameter **p**, that is, a word **w** where each element of $\{0,1\}^p$ occurs just once as a factor of **w**. The number of states of the suffix automaton of **w** can be theoretically computed as being $2^{p+1} - 1$ [10]. We can build this automaton using **suffaut** or **reduce**, and compare the results:

```
$ good 7 | suffaut
word length: 134
states: 255 (8 finals) edges: 381
```

```
$ good -s 7 | reduce
states: 255 (8 finals) edges: 381
```

In order to generate the word, **good** uses the classical construction based on Eulerian Tours. When called with the -s switch, it generates the word and all its suffixes. This script is available with the **biba** distribution.

In the next two examples **biba** deals with natural language dictionaries and biological vocabularies. In both cases we're building reduced automata, in the first case for the entire **br.ispell** dictionary for brazilian portuguese [11] (just like done by Lucchesi and Kowaltowski in [5]), which contains 195539 entries (total size 2120562), and, in the second, for the blocks protein database [4] (109660 entries, 3249189 total size). The filter **eb**, which extracts the sequences from the database distribution file removing comments, is provided in the **biba** distribution.

```
$ ispell -e vocab -d br | reduce
states: 14119 edges: 36588
ram pages: 1300 virtual pages: 1303
```

```
$ eb | reduce -s 200000 -m 28 -c 28 -d 1000000
states: 1318407, edges: 1401826
ram pages: 6200 virtual pages: 6182
```

The first command took just 240 seconds to complete in the same 486 computer described. The second took 27 minutes. Note that this time may vary largely depending on the choice of the paging parameters. We're omitting the final size of the files **a.aut.st** and **a.aut.ed** because in these cases they're full of garbage. In fact, their sizes are not determined by the automaton size, but by the vocabulary size, because the algorithm first builds a trie (this behaviour may be partially avoided using a command-line option). Anyway, at the end of the construction, each state uses 9 bytes, and each edge uses 5 bytes.

The library can also be used for many other purposes, but develop and link programs with the **biba** library is a time-consuming exercise of **C** programming, so we offer a simplistic alternative that opens partially the library's internal functionality for scripts that can be quickly written using **perl** or similar languages.

One trivial script is available in the distribution. It reads the file specified as its first parameter, reads each line, checks if it is or not spelled by a given automaton, and prints it in the affirmative case. Hence, this script can be used as a filter. A more elaborate example is also contained in the **biba** distribution. It computes the *complexity* of a biological sequence as define by Trifonov in [9].

3   Short API Description

For the programmer we provide a short description of the C API (for a detailed description please refer the man pages). Just like we said that automata are like files to the user, they are like files to the programmer too. So deal with automata using this API is like deal with files through the C language library. Automata handles are provided, and they can be used to create or specify an existing automaton, and to subsequently work on it.

create_aut   similar to **create(2)**
register_aut similar to **open(2)**
close_aut   similar to **close(2)**

In order to deal with automata states and edges, very simple services are provided. Each one require the specification of a handle, amongst other specific parameters. The complete list follows:

new_state alloc a new state
set_final   set (to a given value) the state final flag
is_final   read the state final flag
set_dest   set the destination of an edge
dest   read the destination of an edge
stdeg   read the output degree of a state
next_dest visit all edges leaving a given state
clone   clone a state

For low level configuration and session recovery support, the following services are provided:

init_aut   initializes the API and specify paging parameters
stop_pgout disallow page outs, in order to keep
flush_aut   synchronize the disk files with the contents of RAM memory

4   Internal Representation of Automata

States and edges reside on separate data structures, that grow as needed (currently they cannot shrink). The main goal of these data structures is to save memory, because we want to be able to represent large automata, so we need to try to minimize paging at the expense of a larger cost of in-memory operations.

The data structure that represents states is nothing more than a large array. Each entry of this array is a "state", composed of fixed fields (its output degree, a flag to identify if the state is final, and a pointer to the structure that represents its edges) and application-dependent fields, like flags and counters.

The data structure that represents the edges is an array too. It limits the alphabet size to 256, and currently does not support nondeterministic automata.

In order to represent the edges that leave a state with output degree d, where d is in the range 0..256, it allocates just d contiguous entries of the array of edges. The allocation routines guarantee that these d entries reside on the same page, a fundamental property, needed to avoid performance degradation. Each edge is composed by a label field and a destination field which points to a state. Application-specific fields are not allowed.

To allocate a state is no more than increase by 1 the current size of the array of states, and manage the necessary changes in the data structure, which may include the allocation of a new page. Inserting an edge leaving a state demands the allocation of d contiguous entries of the array of edges (best-fit is used), where d is the resulting output degree. This is managed by the maintenance of free lists of free areas within this array, one per each possible free area size.

Both arrays reside on disk, but at each moment only some of their pages are loaded into memory, accordingly to the paging routines. Each state and each edge uses just 40 bits to be represented (the size of a state may be larger depending on the presence of application-specific fields). This number is due to current limit on the alphabet size, that cannot be larger than 256, plus the choice of a 31-bit addressing limit for states and edges. So if n is the sum of the number of states and edges of an automaton, then the total disk size needed to represent it will be 5n bytes.

Page replacement proceeds swapping out pages based on their ages. The pages are circularly visited along the program execution, just like a memory refresh circuit does. The age of a page is the amount of time elapsed since the last time it was acessed. Comparing the age of a page with the medium age computed over all pages, one can decide if it will be swapped out or not.

5   Limits and Limitations

The limits imposed by addressing strategies or performance requirements follow. We plan to relax some of them in future releases of the library (see the section **Future Directions**).

- The states of one automaton can use a virtual space of at most 8 gigabytes. The same limit applies to its edges (this is imposed by a page addressing limit).
- Paging makes unfeasible to adopt a platform-independent representation of data, so when moving automata from one computer to another, conversion tools may be needed (such tools are not currently available).
- To avoid performance degradation, it is not currently allowed to attach application-specific fields to edges nor to use alphabets with sizes larger than 256.
- Nondeterministic automata are not currently supported.
- Automata splitting in many files is currently limited to 10 parts for the states and 10 parts for the edges. So an automaton may be split in at most 20 files.

6 Notes on Performance

The paging scheme by itself represents a performance penalty, because of the additional memory and time comsumption needed to manage and detect page faults and access memory. In a user-level paging scheme like the one implemented in **biba**, this penalty is more significant because we cannot use, as the operating system kernel does, hardware features that make the paging more efficient (and unportable).

In order to estimate the time penalty, we depend on more programming effort. Using the suffix array tool in development (details in the next section) we'll be able to make it.

On the other hand, regarding the memory usage, paging is not critical. The tables and additional information needed to manage paging by **biba** represent less than 1.3amount of physical memory allocated (details on how this percentage was computed can be found in the file mem.c of the distribution).

7 Future Directions

Regarding edges representation, some enhancements are needed, like support for nondeterministic automata. Changes are being planned in order to allow the addition of application-specific fields to edges, as well as the usage of larger alphabets and/or support automata where the labels of the edges are words over a given alphabet.

A search tool for WWW servers similar to **glimpse** [7] will be released soon. It currently uses an index based on **GNU gdbm**, but an automaton-based alternative is being added.

More general data structures besides automata can be represented using **biba**. We plan to release soon a program that builds the suffix array, and implements both the refinement algorithm due to Manber and Myers [6] and the external sorting algorithm due to Gonnet [3]. We wrote this program some time ago, and it is currently being adapted to use the **biba** paging scheme.

A full access to the internal library capabilities though **perl** scripts is being worked on. We currently do not have a schedule for other automata or graph algorithms to be implemented, but we're accepting suggestions.

8 Acknowledgements

We are thankful to Imre Simon by his continuous advice and encouragement. We must acknowledge our master thesis examinators Nivio Ziviani and Yoshiharu Kohayakawa. The biologists Romeu Guimarães and Sérgio Mattioli pointed out biologic resources. Carlos Juiti Watanabe and Eduardo Ermel Wronowski helped to make the implementation. Thanks to Jean-Eric Pin, Marcelo Oliveira, Marko Loparic and the referees for their suggestions. This work was partly supported by Conselho Nacional de Desenvolvimento Científico e Tecnológico (CNPq).

9   Availability

The sources are available under the terms of the GNU GPL, and can be copied from ftp://ftp.ime.usp.br/pub/ueda/biba. General, programming and last-minute informations are available at http://www.ime.usp.br/~ueda/biba.

References

[1] Maurice J. Bach. The Design of the Unix Operating System. Prentice Hall, 1986.

[2] A. Blumer, J. Blumer, A. Ehrenfeucht, D. Haussler, M. T. Chen and J. Seiferas, The smallest automaton recognizing the subwords of a text. Theoretical Computer Science, vol. 40, pp 31-55, 1985.

[3] Gaston H. Gonnet, Ricardo A. Baeza-Yates and T. Snider. Lexicographical indices for text: inverted files vs. PAT trees. TR OED-91-01, UW Centre for the New Oxford English Dictionary, University of Waterloo, 1991.

[4] S Henikoff and JG Henikoff. Automated assembly of protein blocks for database searching, Nucleic Acids Res. 19:6565-6572, 1991.

[5] Cláudio L. Lucchesi and Tomasz Kowaltowski, Applications of Finite Automata Representing Large Vocabularies, in Software Practice and Experience, Vol. 23, pp. 15-30, 1993.

[6] Udi Manber and Gene Myers. Suffix arrays: A new method for on-line string searches. SIAM Journal on Computing, 22(5):935-948, October 1993.

[7] Udi Manber and Sun Wu, GLIMPSE: A Tool to Search Through Entire File Systems, Usenix Winter 1994 Technical Conference, San Francisco pp. 23-32, 1994.

[8] Dominique Revuz, Minimization of Acyclic Deterministic Automata in Linear Time. Theoretical Computer Science, vol. 92 pp 181-189, 1992.

[9] Edward N. Trifonov, Making sense of the human genome. Structure & Methods, 1:69-77, 1990.

[10] Ricardo Ueda Karpischek. The Suffix Automaton (in portuguese). Master Thesis, University of São Paulo, 1993.

[11] Ricardo Ueda Karpischek, **br.ispell**, a brazilian portuguese dictionary for *ispell*, http://www.ime.usp.br/ ueda/br.ispell/, 1995.

On the Syntax, Semantics, and Implementation of a Graph-Based Computational Environment

Yuri Velinov

Department of Computer Science, UNP
University of Natal
Private Bag X01, Scottsville 3209, Pietermaritzburg, South Africa
yuri@cs.unp.ac.za

Abstract. In the present paper we consider the abstract background for designing a practical graph-based computational environment with variable, optional semantics. From the variety of possibilities we concentrate on graphs and polynets as possible carriers of the syntax, and finite automata and flow-diagram programs as possible semantics. We discuss the encapsulation property which emerges in such systems and give precise description of the syntax, operational and denotational semantics in terms of Category Theory. A data structure capable to meet the requirements of a graph-based computational environment is sketched at the end.

Graph-based representations are a usual vehicle for describing processes or events in contemporary science. In particular, they can be used to describe computational processes. Finite automata were possibly the first attempt in this direction. The flow-diagram descriptions of algorithms are another typical example, as ancient as the programming itself. Finite automata and flow-diagrams attracted the attention of the scientists because of their simplicity and transparency, and were subject of active theoretical research. Nevertheless, this research did not result in a practical graphical computational environment for the absence of appropriate substantial medium which can comfortably accommodate something more than the elementary cases. With the contemporary advanced technology and its ability to manipulate visual information in real time there is an actual possibility for implementation of such an environment and graph-based computational languages. Computational languages of this type fall in the class of the visual programming languages. In the present paper we analyze the theoretical background of their possible syntax, semantics and implementation.

The subject of discussion in the further exposition is an environment for geometrical, interactive manipulation of graphs which under appropriate restrictions on the form can carry as options a variety of possible semantics and implement computations prescribed by the currently selected one.

We will restrict our attention to the two most natural candidates for possible semantics - finite automata and flow-diagram programs. Finite automata in all their different forms are a well developed concept naturally connected with graphs. On the other hand flow-diagram programs are the most transparent representation of the structure of algorithms. In a sense, flow-diagram programs are another form of representation of finite automata more appropriate for the task of programming, where the permanent tracking of the global state of the system is an unnecessary burden.

1   Alternatives and Their Formal Background

Any particular program or automaton in the considered environment has two sides - an external appearance, which is a geometrical picture on appropriate medium (say the display of a computer), and its internal representation. The link between these sides is an abstract structure similar to graphs. Both the geometrical and the internal representation in any particular case have as a core an isomprphic image of an instance of such a structure. Therefore, an abstract mathematical structure is underlying any representation. In this section we will discus briefly several candidates which can be used as a basis. The ordinary directed graphs are the most natural candidate and a good starting point for our analysis.

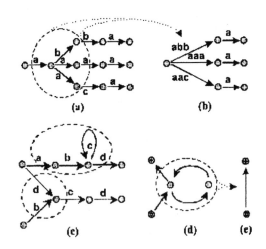

Fig. 1.

We consider a **directed graph** *as a tuple*

$$\mathbf{G} = \langle N, A, sor, tar \rangle$$

consisting of a set of nodes $\mathsf{Nd}(\mathbf{G}) \doteq N$, *a set of arrows* $\mathsf{Ar}(\mathbf{G}) \doteq A$, *and two functions* $sor, tar : A \to N$ *which put in correspondence to each arrow its source and target nodes respectively. This understanding of graphs has the advantage* that it permits consideration of several arrows in the same direction between two nodes. It can also be directly embedded in the framework of Category Theory for resolving the semantic problems. A finite graph of appropriately restricted form can be considered as a representation of a computational process. The imposed restrictions depend on the intended semantics.

If a graph is to be interpreted as a deterministic finite automaton of Mealy type the nodes must be labeled with states, the arrows with pairs of input/output signals (see Fig.1a,c considering a,b,c and d as names of such pairs). Further-more, appropriate rules for the flow of the control must be imposed. Additional restrictions may require that the arrows outgoing from a node must be labeled with different input signals (determinism) and that there are enough outgoing arrows for each node to cover each input signal (totality). The sematic goals of an interpreted graph can be described on abstract level using appropriate category associated with it.

If a graph is to be interpreted as a flow-diagram, according to the traditional flow-diagram approach, the nodes must represent computational steps and the arrows must represent transitions from one computational step to another i.e. the control. Typically, the nodes of the graph are divided into two groups: nodes for actions and nodes for decisions. Only one arrow can originate at an action node and usually only two arrows can originate at a decision node. A node is specified as the beginning of the prescribed process and several nodes can be specified as its ends. A general restriction on the form of a flow-diagram (if only deterministic processes are of interest) requires that any two paths from one node to another must branch through a common decision node.

An alternative possibility to the traditional flow-diagrams, advocated by Landin [2], Burstall [1], Goguen [5], is to consider the arrows as computational processes and the nodes as data containers (Fig.1). Except for its finiteness no other restrictions on the form of the graph are necessary for it to represent a computational process. Appropriate restrictions are imposed on a semantic level instead. Arrows with the same source are considered as alternative. Alternative arrows are used to represent decision steps and have a special interpretation. We will give preference to this approach not only because of its connection with Cat-egory Theory but also because any computation, working on data and producing results, creates a sense of direction. In addition, it can also reflect the parallel structure of the algorithms and is related to the typed functional programming.

A bare graph of a system as described above represents the syntax of a program. From the point of view of the geometric image a system which manip-ulates the syntax must be able to create/delete nodes; to create/delete arrows and associate them with the nodes; to select nodes and arrows and to move them to different positions in the medium; to create/delete intermediate points (pseudonodes) for changing the direction or the shape of the arrows. In addition the following encapsulation property is not only desirable but also necessary for the system to be able to give a global, bird-eye view of the picture and make it more observable:

the system must be able to encapsulate (shrink) the objects in an outlined area into an arrow keeping intact the outside structure of the graph,
and also be able to restore the encapsulated arrow back to the original
form which it represents.

In essence, this means the ability to shrink a subgraph to an arrow (correlate
Fig.1a and Fig.1b, Fig.1c and Fig.2a, Fig.2c, and Fig.2d). The ability of the
system to meet such a property depends strongly on the intended semantics
and cannot be fulfilled completely. Intuitively an outlined area in the case of
automata interpretation may select something as a "subautomaton" (not in the
algebraic sense of this term) or in the case of a flow-diagrams - a subprogram.
Forgetting the semantics for a moment, even on a purely syntactical level, the ordinary graphs can meet this requirement in a very limited form: only an isolated
chain of arrows can be represented in compressed form as one arrow. Otherwise,
multi-head multi-tail arrows must be introduced (Fig.2). This leads to the conclusion that in general another more rich mathematical structure should be taken
as an abstract basis instead of the ordinary graphs. We will consider polynets
and polygraphs (generalizations of graphs where the arrows are multi-tail, multi-
head dragons) as an alternative possibility.

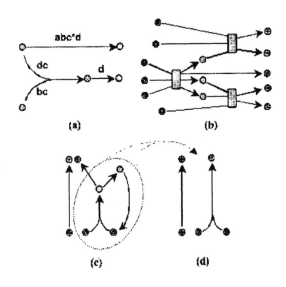

$$\text{Fig. 2.}$$

A polynet

$$\mathbf{N} = \langle N, A, sor, tar \rangle$$

consist of a set of nodes $\mathsf{Nd}(\mathbf{N}) \doteq N$, a set of polyarrows $\mathsf{Ar}(\mathbf{N}) \doteq A$ and two
functions $sor : A \to \mathcal{P}(N)$ and $tar : A \to \mathcal{P}(N)$. The functions sor and tar put

in correspondence sets of nodes to each polyarrow. Thus a polyarrow can have several source nodes and several target nodes.

A structure which can be used as an intermediate step for categorical treatment of the semantics of polynets are polygraphs [7][8]. Similarly to graphs, *a* **polygraph**

$$\langle O, A, +, \Lambda, dom, cod \rangle$$

consists of node-objects O, arrows A and two functions dom, cod : A → O which put in correspondence one source object and one target object to each arrow. On the other hand the objects have structure: together with the operation + : A×A → A and the unit object Λ they form a monoid. Additional requirements can be imposed on the monoidal structure to be able to treat more specific cases. For example, Meseguer and Montanari [6] consider commutative monoids to describe processes in Petri Nets. In the cases appropriate for our purpose the monoid of objects is the free monoid generated by a set of nodes N. Thus an object is a tuple of nodes. Under this restriction polygraphs resemble polynets in that each arrow can have many nodes as sources and targets but in a polygraph the source nodes or the target nodes of an arrow are organized as tuples and because of this repetitions are possible. In essence, this means that different nodes can be interpreted to range over the same domain. A finite polynet in which the nodes are linearly ordered can be considered as a generator of a polygraph. The monoid of the generated polygraph is the free monoid over the nodes. The arrows stay the same but their sources and targets are determined as tuples taking into account the order imposed on the nodes. Though favorable from the point of view of the semantics, polygraphs are not convenient for geometric representation even only for the reason that the free monoid, containing an infinite number of elements, cannot be displayed. This is why it is better to consider them only as an intermediate step to the semantic constructions.

The semantic constructs for ordinary graph-programs can again be given in the framework of Category Theory.

A category is a graph with an additional binary partial operation on arrows subject to specific axioms. Formally, *a* **category** *is a tuple*

$$\langle O, A, \circ, dom, cod, I \rangle$$

where $\langle O, A, dom, cod \rangle$ is a graph with a set A of arrows and a set O of nodes called objects, $\circ : A \times A \to A$ is a partial operation on arrows, and $I : O \to A$ is a function which associates an arrow called identity arrow with each object. The components of a category fulfil the following axioms:

- *The composite $x \circ y$ of any two arrows x and y exists iff $dom(x) = cod(y)$, and in this case $dom(x \circ y) = dom(y)$ and $cod(x \circ y) = cod(x)$;*
- *For any three arrows x, y, z, $(x \circ y) \circ z = x \circ (y \circ z)$ provided the described composites exist;*
- *For any arrow x, $x \circ I(dom(x)) = I(cod(x)) \circ x = x$.*

More information about categories can be found in [5][4]. It is easy to recognize the underlying graph structure of any category correlating the functions *dom* and *cod* with *sor* and *tar* respectively. A typical example of a category is the category $\mathbb{S}ET$ which has sets as objects and total functions as arrows. We will use the category $\mathbb{C}FN$ of the computable functions, which has constructive domains as objects (domains generated from a finite initial set by several construction operations) and computable functions as arrows.

Polycategories are similar to categories but have additional monoidal structure on objects and an operation on arrows which includes objects as a necessary specification. Formally, *a* Y–**polycategory** *is a tuple*

$$\langle O, A, +, \Lambda, dom, cod, \gamma, I \rangle$$

where $\langle O, A, +, \Lambda, dom, cod \rangle$ *is a polygraph,* $I : O \to A$ *and* $\gamma : A \times O \times O \times O \times A \to A$ *is the arrow operation which, when defined, puts in correspondence to each tuple* $\langle x, a, b, c, y \rangle$ *an arrow denoted as* $x\lceil a, b, c \rceil y$ *(the triple of objects in this notation specifies the way the two arrows* x *and* y *are composed). The components of a polycategory fulfill several axioms as follows.*
For any arrows $x, y, z,$ *and objects* a, b, c, d, e :

- $x\lceil a, b, c \rceil y$ *exists iff* $dom(x) = a + b + c$ *and* $cod(y) = b$;
- *if* $x\lceil a, b, c \rceil y$ *exists then* $cod(x\lceil a, b, c \rceil y) = cod(x)$, $dom(x\lceil a, b, c \rceil y) = a + dom(y) + b$;
- $(x\lceil a, b, c + d + e \rceil y)\lceil a + dom(y) + c, d, e \rceil z = (x\lceil a + b + c, d, e \rceil z)\lceil a, b, c + dom(z) + e \rceil y$ *provided the described composites exist* (**commutativity***);*
- $(x\lceil a, b, c \rceil y)\lceil a + d, cod(z), e + c \rceil z = x\lceil a, b, c \rceil (y\lceil d, cod(z), e \rceil z)$ *provided the described composites exist (which requires that* $b = d + cod(z) + e)$ (**associativity***);*
- $I(cod(x))\lceil \Lambda, cod(x), \Lambda \rceil x = x\lceil \Lambda, dom(x), \Lambda \rceil I(dom(x)) = x$

Fig.3 illustrates some of the axioms.

A typical example of a polycategory is the polycategory $\mathbb{P}FN$ which has sets as objects and functions as arrows and where the monoidal operation is a slightly modified (because of Λ) Cartesian product. We will use the polycategory $\mathbb{P}CF$ of computable functions which has constructive domains as objects and computable functions as arrows. More information about polycategories can be found in [7][8].

Structures of the type considered above can be related by morphisms. *A* **graph morphism** $F : \mathbf{G}_1 \to \mathbf{G}_2$ *between two graphs* $\mathbf{G}_1$ *and* $\mathbf{G}_2$, *is a pair of functions*

$$\langle F_\bullet : \mathsf{Nd}(\mathbf{G}_1) \to \mathsf{Nd}(\mathbf{G}_2), \; F_\blacktriangle : \mathsf{Ar}(\mathbf{G}_1) \to \mathsf{Ar}(\mathbf{G}_2) \rangle,$$

which preserve the sources and the targets of the arrows i.e. for any arrow $x \in \mathsf{Ar}(\mathbf{G}_1),$ $sor(F_\blacktriangle(x)) = F_\bullet(sor(x))$ *and* $tar(F_\blacktriangle(x)) = F_\bullet(tar(x))$. *Category morphisms are called* **functors**. A functor is a graph morphism on the graph

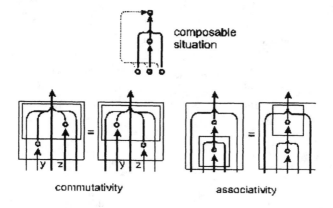

Fig. 3.

parts of the related categories and in addition it preserves the arrow composition operation and the identity arrows. The precise definition of a functor can be found in [5][4]. The definition of a Y-polycategory functor is given in [7][8].

If polynets are used as an abstract basis the programming environment becomes more flexible and is able to accommodate other possible interpretations. For example, multi-input/output automata can be considered as building blocks of more complex systems obtained by connecting together some of their input and output lines (Fig.2b). In this case (as it is with the flow-diagrams) nodes keeps the input/output signals. Logical circuits or Neural Nets are specific cases of such interpretation. Systems of Petri-nets type used for describing computational processes can be a source of other interpretations. Just to underline how rich the area of possible interpretations really is remember that even the understanding of automata can be different - they can be of different types, say Mealy or Moore, and they can also be considered as transducers or as acceptors; a different flavor may be, but one which requires different interpretations and semantic treatment.

2   Abstract Structure of the Graph-Based Programming Languages

The analysis carried out above gives enough background to outline the structure of some graph-based programming languages. A thorough exposition of the variety of feasible interpretations is impossible in a small paper. In the present section we will give a more thorough description of the syntactical and semantic sides of a computational environment for automata and flow-diagrams. From the alternative possibilities considered in the previous section we select the ordinary graphs and the polynets, intending the arrows to represent computations.

Any finite graph can be considered as a graph-automaton scheme. In order to describe the automata interpretations of such a scheme it is convenient, given two sets of signals X and Y, to denote by $(X \times Y)^{\clubsuit}$ the subset of $X^* \times Y^*$ consisting of all pairs of strings of equal length (notice that there is a natural isomorphism between $(X \times Y)^{\clubsuit}$ and $(X \times Y)^*$ which makes it possible to consider strings of pairs instead of pairs of strings whenever it is more convenient). Having in mind a slightly generalized version of deterministic finite automata of Mealy type, necessary to accommodate a limited form of encapsulation, we can give the following definition.

Definition 1 *A* **graph-automaton** $\langle G, X, Y, \gamma \rangle$ *consists of a finite graph* G, *two finite sets - X of input signals and Y of output signals, and a labeling function* $\gamma : \mathsf{Ar}(G) \to (X \times Y)^{\clubsuit}$.

A graph-automaton is **elementary** *if the codomain of γ is restricted to $X \times Y$.*

An elementary graph-automaton is **total** *iff the number of the outgoing arrows of each node labeled with different input signals is $Card(X)$.*

An elementary graph-automaton is **deterministic** *if (but not only if) all arrows outgoing from a node are labeled with different input signals.*

In this paper our interest is restricted to the deterministic automata. However, we restrain from presenting the cumbersome conditions for determinism in the general case. Instead, we will consider only automata which originate from deterministic elementary automata by encapsulating sequences of arrows as described later.

With each graph automaton $\langle G, X, Y, \gamma \rangle$ the category of paths over G is associated and naturally interpreted. In this category every two arrows $o_0 \xrightarrow{x_1/y_1} o_1$, $o_1 \xrightarrow{x_2/y_2} o_2$ such that the target of the first coincides with the source of the second can be composed to produce the arrow $o_0 \xrightarrow{x_1 x_2/y_1 y_2} o_2$. Identifying the arrows of the category with their labels, the semantics of a graph-automaton can be determined denotationally as a family of functions $(f_o : X^* \to Y^*)$ associated with the nodes of the graph. For each node o

$$f_o(x_1 x_2 ... x_n) = \begin{cases} \text{the second component of the arrow } x_1 x_2 ... x_n / y_1 y_2 ... y_n \\ \qquad\qquad \text{whose source is } o. \end{cases}$$

(this family is nothing more than a form of representation of the well known extended output function of an automaton).

The operational description of the semantics is obvious: trace the graph following the arrows, according to their labels, in the order of the input signals and output the second signal of the label of each passed arrow.

A graph-automaton of the considered form permits encapsulation only of an area with one incoming arrow and several outgoing arrows which comprises a tree subgraph (the internal leaves, if any, are always considered as targets of outgoing arrows). The encapsulation in such a case results in a bunch of arrows with the same source and different targets. Each arrow is the composite of the original arrows in a path which starts with the unique incoming arrow and finishes with an outgoing of the area arrow.

A more general encapsulation property encounters difficulties. Parallelism can be treated in the framework of the partially additive categories [3] but Category Theory does not have appropriate mechanisms to handle loops. An additional unary operation (star operation) can be introduced for this purpose and this may lead to an enriched category structure capable of handling regular expressions. Still the situation will not be quite satisfactory since the correspondence between regular expressions and automata is not one to one. The introduction of **net-automata** where multi-tail-single-head (or single-tail-multi-head) arrows are possible may also be interesting. In general, this direction is not well developed and requires further study.

Aiming to interpret graphs as prescriptions for computational processes we may follow the ideas of Burstall [1] and Goguen [5].

Definition 2 *A* **graph-program scheme** *$\langle G, i, e \rangle$ is a finite graph* **G** *where two (not necessarily different) nodes* **i** *and* **e** *are selected as an initial node and a final node respectively and where every node is reachable from the initial node.*

A simple example of a graph program scheme can be seen in 1b where the initial node is marked with ∘ and the final node is marked with +).

An interpreted program scheme becomes a graph-program. To interpret a program scheme $\langle G, i, e \rangle$ we map it by a graph-morphism into the category of computable functions fulfilling the additional restriction that the alternative arrows outgoing from a node must have as images functions with disjoint domains of definition. Thus, formally:

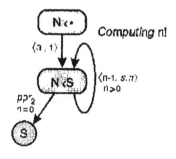

Fig. 4.

Definition 3 *A* **graph-program** *is tuple* $\mathbf{P} = \langle \mathbf{G}, \mathbf{i}, \mathbf{e}, F \rangle$ *where* $\langle \mathbf{G}, \mathbf{i}, \mathbf{e} \rangle$ *is a graph program scheme and* $F : \mathbf{G} \rightarrow \mathbb{C}FN$ *is graph-morphism which puts in correspondence constructive domains to the nodes and computable functions to the arrows, subject to the restriction that the functions associated with alternative arrows have disjoint domains of definition.*

Fig.4 shows a graph program for computing of $n!$.

The semantics of a graph-program **P** is a computable function. It can be determined in operational manner as follows:

- Associate with each node a variable ranging over its corresponding set (or a tuple of variables if the corresponding set is a product);
- Assign initial values to the variables of the initial node;
- In all possible cases consume the values assigned to the variables of a node by computing the function which has it as a source and assigning the value of the function to the variables of its target node. In any case of alternative arrows choose the one which represents function defined for the currently available value in its domain;
- Continue the above process until the final node is reached. The obtained value from the set associated with the final node is the value of the function determined by the program.

Notice that with this approach we may think that a function is sensing its domain for availability of data (compare this observation with the token game for net-programs described below).

The semantics of a graph-program $\mathbf{P} = \langle \mathbf{G}, \mathbf{i}, \mathbf{e}, F \rangle$ can be given in denotational manner using the technique of Category Theory. The graph **G** of the program generates a free category $\mathbf{G}^*$ - the category of paths in **G** (the arrow operation simply joins two connected paths together). The graph morphism $F : \mathbf{G} \to \mathbb{C}FN$ can be extended in a unique way to a category functor $F^* : \mathbf{G}^* \to \mathbb{C}FN$. It can be proved that the union of all functions corresponding to the arrows in $\mathbf{G}^*$ with source **i** and target **e** is a function. This function is the one computed by **P** i.e.

$$\text{the meaning of } \mathbf{P} = \bigcup_{p \in Hom_{\mathbf{G}^*}(\mathbf{i}, \mathbf{e})} F^*(p) \ .$$

A comprehensive description of the above construction can be found in [5].

From the discussion above it is easy to see that any area with one incoming and one outgoing arrow comprises itself a graph-program (a subprogram of the original one). Therefore, any such area can be encapsulated to an arrow and supplied with the corresponding meaning (see Fig.1d,e). Notice that the initial and the final nodes are always considered as external to any area.

Polynets together with polygraphs used as a foundation can give more expressive power and flexibility to a graphical computational environment. In particular they inherently incorporate parallelism.

Definition 4 *A* **net-program scheme** $\langle \mathbf{N}, S, T \rangle$ *is a polynet* **N** *with distinguished a set of initial nodes S and a set of final nodes T such that:*

- *every node is reachable from at least one initial node;*
- *every node which is not a target of an arrow is an initial node and every node which is not a source of an arrow is a target node;*

- *for every two paths which have common target nodes one is part of the other or they are alternative (in the sense that they have a common initial part which might be empty and then split through alternative arrows).*

Fig.2e shows a simple net-program scheme.

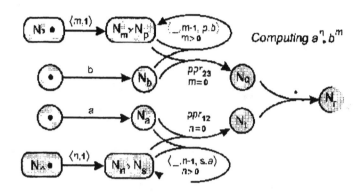

Fig. 5.

To interpret a net program scheme $\langle N, S, T \rangle$ it is necessary to accommodate its arrows to the structure of the multidomain functions which are usually defined on products. For this purpose we first associate a polygraph N° with the net-program scheme by considering the free monoid of its nodes and then by associating objects of the free monoid with the arrows as their sources or targets and also with the initial nodes and the final nodes. This must be done in such a way that the domain or codomain of an arrow is a tuple of its source or target nodes respectively. Then, the initial object S° of N° is a tuple of the nodes in S and the final object T° of N° is a tuple of the nodes in T. Such association can be done in different ways. Once N° is obtained an interpretation of $\langle N, S, T \rangle$ can be considered as a polygraph morphism $F : N^\circ \to \mathbb{P}CF$ from the polygraph N° to the polycategory of computable functions, such that the functions which correspond to alternative arrows have different domains of definition on the subdomains of their common nodes. Thus formally:

Definition 5 *A **net-program** is a tuple $\langle N^\circ, S^\circ, T^\circ, F \rangle$ where S°, T° are objects of N° (the initial and the final objects respectively) and $F : N^\circ \to \mathbb{P}CF$ is a polygraph morphism to the polycategory of computable functions fulfilling the "alternative arrows requirement".*

Fig.5 shows a simple net-program for computing of $a^n b^m$.

The operational semantics of a net program can be given at this stage by following some rules for the control of the computational process. These rules can be described as a token game which though similar in spirit is quite different from the token game used in Petri Nets Theory. We distinguish two types of tokens - active and passive. A node can contain at most one token.

- At the beginning (step 0) all initial nodes contain active tokens;
- At step t each arrow not alternative to another, and such that all its source nodes contain a token at least one of which is active, is activated. In alternative situations just one of the possible alternative arrows which fulfil the above condition is activated. An activated arrow changes its source tokens to passive and produces active tokens in its target nodes.
- The token flow stops when there is no possibility for it to be continued.

A computation goes along with the above rules by computing at each step the values of the functions corresponding to the activated arrows for the available data. In case of alternative arrows the one which represents the function defined for the available data must be chosen.

The denotational semantics of a net program can be given in categorical manner if the underlying net fulfills additional restrictions necessary to accommodate it to the expressive abilities of the arrow composition operation in a polycategory. The transition from $\mathbf{N}$ to $\mathbf{N}^\circ$ must keep completely all aspects of the connectivity of the arrows of $\mathbf{N}$. In essence, it must be such that

- the common sources (targets) of any two arrows form a subtuple of their domains or codomains respectively,
- the common sources (targets) of any arrow with the initial or final object form its subtuple respectively.

For example, these requirements can be met if $\mathbf{N}$ is planar. Under this condition the semantics of a net program $\mathbf{P} = \langle \mathbf{N}, S^\circ, T^\circ, F \rangle$ can be given in a way similar to the one considered for graph programs.

$$\text{the meaning of } \mathbf{P} = \bigcup_{p \in Hom_{\mathbf{N}^*}(S^\circ, T^\circ)} F^*(p)$$

where p are arrows in the polycategory $\mathbf{N}^*$ freely generated by $\mathbf{N}^\circ$ (see [7] for details) and S°, T° are the initial and the final object. A more thorough description of this construction can be found in [8].

The encapsulation resources of the net-programs are much better than those of the other considered cases. The definitions of a net-program scheme and a net-program show immediately that any area with one outgoing arrow (initial and final nodes are always considered as external to any area) comprises a net program and therefore can be encapsulated to an arrow associated with appropriate semantics (Fig.2c,d).

3   Remarks on the Implementation

The discussion in the previous sections gives enough formal background for practical design of a graph-based computational environment.

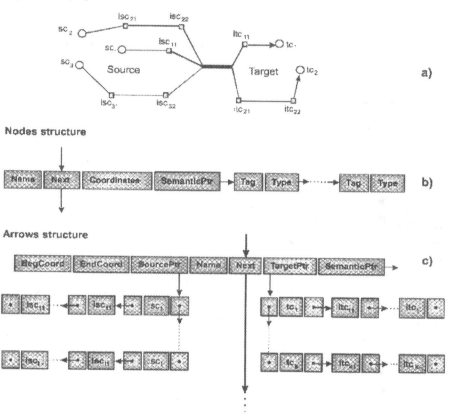

Fig. 6.

Since graphs can be considered as a partial case of polynets, for universal-ity, we prefer the last as a syntactical basis. A data structure for the internal implementation of the environment on a conventional computer must embrace si-multaneously the parameters of the geometric picture (which are not important for the computational process) and the graph structure. The standard imple-mentations of graphs are not appropriate for our purpose. Suppose that in its most coarse form an arrow (Fig.6a) is pictured as several polylines attached to a geometric shape used as a core. A simple shape of the core can be a line segment (the thick line in the picture). Then, a possible data structure which can incor-porate the desired information is represented in Fig.6b,c. It consists of a linked list of nodes and a linked list of arrows. Each node, in addition to its name and coordinates, carries semantic information in the form of a tag-type linked list (we allow a node to represent a tuple of domains). Each arrow contains information about its sources and targets recognized by their coordinates. It also contains information about the parameters of its geometric core and coordinates of the

intermediate points for the outgoing or incoming geometric arrows in the form
of properly organized linked lists. A semantic pointer is supposed to associate
proper computations with an arrow. The fields which are not necessary under
some of the interpretations (like the semantic fields of the nodes in the Mealy
automata case) are ignored.

4   Conclusions

In the previous sections we have considered some possibilities for designing of
a graph-based computational environment. It can be seen that polynets are the
necessary formal basis if a more complete encapsulation property is desired. The
creation of a real graph-based programming language is in the favorable situation of existence of solid theoretical foundations and technological premisses.
Nevertheless, one can hardly expect that a graphical programming language,
the way it is considered here, can be a substitute for an ordinary programming
language. If the arrows represent primitive operations the graph of a serious
program will be too big to be comfortably observed. In addition a proper mechanism for recursive definitions is still missing. On the other hand, the graphical
approach can give a better description (from the human point of view) of the
global structure of a program. Therefore, it is more realistic to consider and
design graph-based programming languages as extensions to the conventional
programming languages. The arrows can represent complex computations described in a conventional manner and a graph program can generate an ordinary
program which can be processed further by an ordinary compiler. On the other
hand graph based computational languages can be very convenient and fruitful
in areas where recursion mechanisms are not necessary.

References

1. Burstall, R. An Algebraic Description of Programs with Assertions, Verification and
   Simulation, Proceedings of ACM Conference on Proving Assertions about Programs,
   Las Cruces, New Mexico pp.7-14,(1972).
2. Landin, P.J. A program machine symmetric automata theory. Machine Intelligence
   5 (eds. B. Meltzer and D. Michie) Edinburgh University Press, pp.99-120,(1969).
3. Manes, E., M. Arbib. Algebraic Approaches to Program Semantics, Springer-Verlag,
   (1986)
4. Mac Lane, S. Categories for Working Mathematicians, Springer-Verlag, (1971).
5. Goguen, J., On Homomorphisms, Correctness, Termination, Unfoldment and Equivalence of Flow Diagram Programs, Journal of Computer and System Sciences 8,
   pp.333-365 (1974).
6. Meseguer, J., U. Montanari. Petri Nets are Monoids, Tech. Rep. SRI-CSL-88-3,
   C.S.Lab.,SRI International, January (1988).
7. Velinov, Y. An algebraic Structure for Derivations in Rewriting Systems, TCS 57,
   pp.205-224, (1988).
8. Velinov, Y. Nets Polycategories and Semantics of Parallel Programs, X National
   School for Scientists "Applications of Math. in Techniques", Varna, pp.340-351,
   (1985).

The Finite State Automata's Design Patterns

Sandro Pedrazzini

[1] IFI-WWZ
University of Basel
Petersgraben 51, 4051 Basel, Switzerland
[2] IDSIA
Corso Elvezia 36, 6900 Lugano, Switzerland
sandro@idsia.ch

Abstract. In this article we want to discuss the design patterns used and proposed for the realization of finite state automata. Various aspects in the design of a framework for the implementation of FSA will be treated, presenting not only the patterns for the single components, but the entire system design. Using design patterns to sketch a framework means performing an "abstract implementation", from which it is possible to realize concrete specific automata, simply customizing some classes. In order to test the framework, some concrete lexical tools have been created. The resulting automata and transducers are used to perform word form analysis, word form generation, creation and derivation history, spellchecking and phrase recognition.

1 Introduction

Finite-state machines and transducers are used in many domains. They allow to perform many operations in an efficient and elegant way. Recently, a number of general-purpose class libraries have been developed at universities and companies ([7]). Purpose of this article is not to present a new toolkit providing implementations of all the known algorithms to optimize the construction and the use of automata. Our purpose is instead to propose and discuss a general design to be used while implementing finite-state automata and extensions of them, such as transducers. From this general object-oriented design, which makes use of object-oriented concepts, without introducing any dependency from any programming language, we have realized a general-purpose framework for finite-state systems implementation, written in C++. Customizing the framework has brought to the realization of different lexical automata. In a certain way the realization of the object-oriented framework has been used as test for the overall design that we will present in this article, whereas the single concrete lexical automata have represented a test for the framework's customization. First of all we must point out what the meaning of a framework is. According to [3], a framework is more than a simple toolkit. It is a set of collaborating classes that make up a reusable design for a specific class of software. The purpose of the framework is to define the overall structure of an application, its partitioning into classes and objects,

J.-M. Champarnaud, D. Maurel, D. Ziadi (Eds.): WIA'98, LNCS 1660, pp. 213-219, 1999

their collaboration, and, most important, the thread of control. These predefined design parameters allows the programmer to concentrate on the specifics of his application. He will customize the framework for a particular application by creating application specific subclasses of classes (eventually abstract) from the framework. The framework itself can be viewed as an abstract finite-state element. Only the definition of some concrete classes can generate from it a usable automaton. The main design decisions have therefore already been taken, and the applications (finite-state elements) are faster to implement and even easier to maintain. Even though the framework has been fully implemented ([5]) and its customization has allowed the realization of many different lexical tools ([6]), also available as client/server through a Web page, the aim of this article is its discussion from a software design point of view, in order to allow reuse from a higher point of view. There are five main components of the framework which will have to be customized. For each of those five parts we have used an existent design pattern ([1]). We also suggest a pattern for a sixth part (external input), which, however, we have not implemented as customizable. For the input/output component we have used the Adapter, for the FSA nodes we have used the Template, for the traversing algorithm the Strategy, for the action to be performed at each node a simplified version of Visitor and, finally, to locally manage the instantiation of each single customizable element, we have used the Factory design pattern.

2   Input/Output: Adapter

The internal structure of a concrete FSA can be printed in different forms. We use a text readable form, when we want to have the opportunity to edit it, in order to understand the internal elements, data types and links of the realized automaton, to eventually allow debugging. We use a crypted form when we want to give our tools for public use. These two kinds of output (and corresponding input) treat the data in different manners and have different interfaces. What we need is a common adapted interface which allows the replacement of the concrete input/output device (object) at any time and without consequences for the rest of the program. We need a wrapper. As sketched in fig. 1, the Adapter design pattern helps realizing a wrapper converting the interface of a class into another interface clients expect. The framework must foresee a mechanism to allow the adaptation of the output, without breaking the interface used.

The new input/output will be wrapped into a new class, which not only performs a call further to the corresponding method of the wrapped object, but also executes some adapting operations on the data. The tasks to be performed in our crypted input/output class are very different from the ones used in the textual output. The adapting operations of the former has caused the implementation of a circular buffer used to store a certain amount of temporary data, waiting, in case of output, to be transformed into the desired crypted data, and, in case of input, to be loaded into the internal structure.

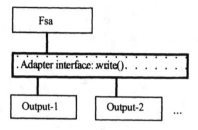

Fig. 1. Adapter interface

3   Node: Template

Each finite-state tool can have a different kind of node, depending on the kind of information it must code and on its use, unidirectional or bi-directional. The opportunity to define a new node represents therefore a level of customization. The new kind of node should take advantage of the existing managing algorithms, using them as they are, without further modifications. There are two methods for realizing such a design: parameterized types and common classes. Parameterized types, even though in general more efficient, is however a concept which is not known in every programming language, and this would restrict the generalization of our software design, which is intended to be independent of any programming language. Because of this and because we have judged the second method more flexible, we have chosen the second one. Notice that the method used is called Template in the design patterns terminology. This is not to be confused with the C++ template mechanism, which represents the C++ way of realizing the parameterized types. The abstract Node class (fig. 2) must define the interface, previewing all basic functionalities required for the nodes by the internal algorithms. The latter will use the concrete elements through Node references.

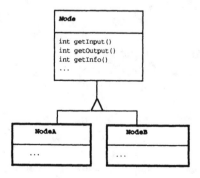

Fig. 2. Node hierarchy

4   Traversing the FSA: Strategy

A transducer used to generate word forms is not deterministic and needs to be traversed in a non deterministic way, an automaton which is known a priori to be deterministic, could choose to use an easier traversing algorithm, which do not looks for all search paths. The looping over nodes is also an opportunity that could be switched off, for some tools, for efficiency purposes. As we can see there are many algorithms to consider for traversing finite-state elements. Hard-wiring all of them into the class that may require them is not desirable. First because the class will get more complex if it has to include all possible algorithms, and different algorithms will be appropriate at different times; second because it will become difficult to add new algorithms or to vary existing ones when traversal is an integral part of the class that uses it. We can avoid these problems encapsulating all different traversing algorithms in different classes, using the same interface, as shown in fig. 3. The interface is defined by a common superclass, the Strategy class. The intent of the Strategy pattern is to define a family of algorithms, encapsulate each one, and make them interchangeable.

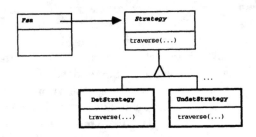

Fig. 3. Strategy

5   Get Information: Visitor

The main feature here is the separation of information extraction performed during the traversal process, from the traversal algorithm itself. We must keep the responsibility of the action away from the traversal part. In this way we can use the same finite-state tool to deliver a different type of information. We need a callback mechanism able to store intermediate results. We had two models: Command and Visitor to describe our design and implementation. We used the Visitor because it seemed nearer to our purposes, even though we used a simplified version of it, because the visitor is thought for a use with different kinds of nodes at the same time, which is not our case. But the meaning is the same: an external object is used to access and read the data of the internal structure. The information extraction process is embedded into the abstract class Visitor. Subclassing it means reusing the nodes of the finite-state system, building with it a new kind of answer. During the retrieval process the internal data in the nodes remains read-only, i.e. unmodified. The adaptation is in the way the data will be used for the external result. For example, the difference between the information extracted from a lemmatizer and the information extracted from a morphosyntactic analyzer can be coded uniquely distinguishing two different interpretations of the same data, i.e. modifying the action performed during the traversal. The traversal process is responsible for leading the control through the structure, whereas the action, which will be called for each node, involves accumulating information during it. This is particularly useful with lexical transducers, which store input and output information in the nodes. Separating the retrieval process from the internal structure will bring more flexibility and potential for reuse, because different kinds of retrieval often require the same kind of traversal. In addition, we will simplify the task of customizing the retrieval, restricting the modification to the action. The implementation is organized as follows: an object of the class Strategy is responsible to traverse the internal structure and has a reference to a Visitor object. During the traversal, each node of the structure will receive the visit of the instantiated Visitor object. The instance is used for accumulating information, creating the final result of the analysis. The overall pattern is shown in fig. 4. The abstract class Visitor is shown with two (among many possible) inheriting concrete classes.

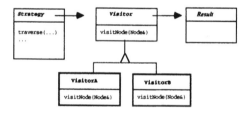

Fig. 4. Visitor

Any FSA internal specific data structure remains separated and hidden for the visitor object, simplifying the task of the customer.

6 Custom Instantiation: Factory

Working with abstract classes to allow customization causes a typical problem during the instantiation of the concrete classes. The framework uses the abstract classes to define and maintain relationships between objects and is also responsible for creating them. But the corresponding concrete classes are defined later, because they are application-specific. The framework must instantiate classes, but it only knows their abstract definition, and cannot predict the concrete subclasses, for example, of Node and Visitor to instantiate. The solution to this dilemma is offered by the virtual constructor, also classified as Factory method in design patterns. This method takes advantage of the fact that the framework does not know the concrete classes to instantiate, but knows exactly when their creators should be applied. It moves out of the framework the responsibility to instantiate concrete classes. The main framework class (in our case the Fsa class) defines the four pure virtual operations createNode(), which will return a Node reference, createVisitor(), which will return a Visitor reference, createStrategy(), which will return a Strategy reference, and createFileWrapper(), which will deliver the right input/output Adapter object. Those operations will be redefined subclassing Fsa. The real implementation of the methods will then be part of the subclass of Fsa, which will instantiate the concrete classes (fig. 5).

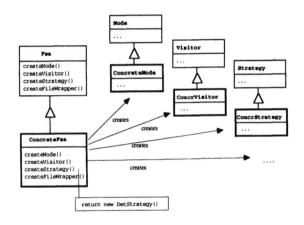

Fig. 5. Factory method

The factory method is good for our system, because it eliminates the need to bind domain-specific classes into the overall framework code, enhancing therefore the customization and the reusability of the framework itself.

7   External Input: Decorator

There is another kind of input, the external input, i.e. the collection of data coming from outside, which must be converted into the internal finite-state structure. Differently from some existing finite-state tools, which need a regular grammar description as input, our framework is based on an extended input, from which it will extract the regular expressions' mechanism to build the automata. We consider as input each single sequence of symbols which has to be recognized and retrieved in the building FSA. Our external input module is at the moment only partly customizable. The system itself is responsible of transforming and optimizing the whole input in a minimal FSA. We know, on the other hand, that other finite-state tools are built having as input other kinds of sequences, regular expressions, continuation classes ([4]), etc. It seems therefore useful to define a filter module for each kind of input, in order to convert it to the type requested by the framework. The conversion could be considered an operation to be applied outside (in this case it would not be part of the overall design) or inside the framework. The easiest way of realizing it inside could be through the use of the Decorator pattern, which would guarantee the complete reuse of the main existent input function and, at the same time, the transparency of the operation.

8   Conclusions

We have shown the software design of a framework representing an abstract finite-state element, which can be easily customized to produce new kinds of concrete finite-state tools. A real framework has been realized (using the C++ language) applying this design, and from this framework various finite-state concrete elements have been implemented. The functionalities of the single elements have been tested and demonstrated ([6]), and can be shown through a client/server demo version on the Web at the following address: http://www.wordmanager.com. But the significance of these tools does not reside primarily in their individual functionalities. Their principal interest lies in the fact that they can be produced with so little effort on the basis of an existing object-oriented customizable framework.

References

1. Gamma E., Helm R., Johnson R., Vlissides J.: Design Patterns, Addison Wesley, 1995.
2. ten Hacken P, Bopp S., Domenig M., Holz D., Hsiung A, Pedrazzini S.: A Knowledge Acquisition and Management System for Morphological Dictionaries, In Proceedings of Coling-94, International Conference on Computer Linguistics, Kyoto, 1994.
3. Johnson R.E and Foote B., Designing reusable classes, Journal of Object-Oriented Programming, June/July 1988.
4. Karttunen Lauri: Finite-State Lexicon Compiler, Xerox Corporation Palo Alto Research Center, Technical Report [P93-00077], 1993.
5. Pedrazzini S.: Applying Software Design for Creating Customizable Lexical Tools, Technical Report 97-2, Universität Basel, July 1997.
6. Pedrazzini S., Hoffmann M.: From Lexical Acquisition to Lexical Reusable Tools, First International Conference on Language Resources and Evaluation (LERC), Granada, Spain, May 1998.
7. Watson B.: Implementing and using finite automata toolkits, in Natural Language Engineering, Vol 2, part 4, pp. 295-302, December 1996.

Automata to Recognize Finite and Infinite Words with at Least Two Factorizations on a Given Finite Set

Xavier Augros

Laboratoire I3S
Université de Nice-Sophia Antipolis
bât. ESSI, BP 145, 06903 Sophia-Antipolis Cedex, France
augros@i3s.unice.fr

Abstract. This paper is about the recognition of finite and infinite words with at least two factorizations on a finite language. We implement the construction of the graph of delays for a finite language L which can be view as an automaton that accept words and ω-words that are ambiguously covered by L.

1 Introduction

The goal of this software is to recognize words which have at least two factorizations on a finite language L for the power-$*$ and the power-ω(see [PP]) of this language. A word α is ambiguously covered by a language L if there is at least two factorizations of α with different first factors in L([Kar85]). To recognize those words, we use a graph inspired by J.Karhumäki in [Kar85] called graph of delays. He used this graph to visualize the two factorizations of an infinite word in the case of a language generated by a code with three elements.

We use graphs of delays to classify different samples of finites generators of ω-languages by the number and the representation of factorizations of words ambiguously covered by the given language. The aim is to decide if, for a given finite language L, there exist a code (or an ω-code) C such that $L^\omega = C^\omega$. We already know that for a code C, such that C^+ is the greatest generator of C^ω, there is no ω-code as ω-generator[JLP96, Jul96] and the graph of delays allows us to decide if a finite language is a code and an ω-code or not.

None of the existing systems[MMP+95, VUMB97] allow us to do this construction, so we have written a software in Java which implements the graph of delays, draws it and uses it as an automaton to recognize words with at least two factorizations on a finite language.

Our program allows us to construct this graph of delays for a finite language and to use it as a non-deterministic automaton to recognize the set of finite and infinite words ambiguously covered by the language L (and then to decide if L is a code or an ω-code). This graph of delays is very close in the spirit to the construction of the test of Sardinas and Paterson to decide if a language is a code[SP53].

J. M. Champarnaud, D. Maurel, D. Ziadi (Eds.): WIA'98, LNCS 1660, pp. 230–235, 1999.

2   Preliminaries

Let Σ be a finite alphabet. The set Σ^* is the set of all finite words over Σ and Σ^ω is the set of all infinite words. The empty word is denoted by ϵ and $\Sigma^+ = \Sigma^* \setminus \{\epsilon\}$. Let $L \subset \Sigma^*$ be a finite language, then L^* is the set of finite words obtained by the finite concatenation of words of L, and L^ω is the set of infinite words (also called ω-words) obtained by infinite concatenation.

A factorization over L of a word $u \in L^*$ is a sequence $(u_1, u_2, \dots, u_k)$ of words of L such as $u = u_1 u_2 \dots u_k$. A factorization over L of a word $\alpha \in L^\omega$ is an infinite sequence $(\alpha_1, \alpha_2, \dots, \alpha_k, \dots)$ of words of L such as $\alpha = \alpha_1 \alpha_2 \dots \alpha_k \dots$.
We say that a word u has two factorizations over a language L (or u is ambiguously covered by L) if u has two factorizations with different first words in L. We denote by $\text{pref}(L)$ (resp. $\text{pref}(u)$) the set of prefixes of words of L (resp. the set of prefixes of the word $u \in L$).
A language C is a code[BP85] iff each finite word of C^* has only one factorization over C, and we say that it is an ω-code iff each infinite word of C^ω has only one factorization over C.
In all the paper, we consider words with two factorizations with none common factors, for example with the language $L = \{a, abac, ba, ca\}$, the word $baabacacaca$ has two factorizations (see Fig.1) but the first factors and the two last ones are the same in both. In such a word, only the factorization of $abaca$ is really ambiguous.

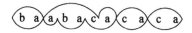

Fig. 1.

3   The Graph of Delay

Given a finite language, we define the graph of delays as an oriented graph whose set of vertices is the set of delays between two factorizations, *i.e.* labels of vertices are advances of one factorization on an other for the prefix of a given ambiguous word. Intuitively a labeled edge fits with the closest prefix of one of the two factorizations from the current read prefix. The delay ϵ means either that we are at the beginning of a word or that we have two factorizations of a finite word. Thus we can decide if a language L is a code and not an ω-*code* , or if this language is an ω-*code* .

Definition 1 *Let L be a language. We call graph of delays the pair (V, E) such that:*

- V *is the set of vertices*
  $V = (L^+)^{-1} L^+ \cap \text{pref}(L)$

- E is the set of oriented edges
  $E = \{(u, l, v) \in V \times \Sigma^+ \times V/(uv \in L \text{ and } l = v) \text{ or } (l \in L, ul = v \in pref(L)) \quad \text{or } (v = \epsilon, l \in L, ul \in L)\}$
  (u, l, v) is an edge labeled by l from a vertex labeled by u to one labeled by v.

The following algorithm construct the graph of delays.
let L be a finite language, V and E are empty set.

- for x and y in L
    - if x is a prefix of y then
        - add x in V
        - add (ϵ, x, x) in E the edge from ϵ to x labeled by x.
- for $v \in V$
    - for p the prefix of a word of L
        - if $vp \in L$
            — if $p \notin V$ then add p in V
            — add (v, p, p) in E the edge from v to p labeled by p.
            — if $p \in L$ then add (v, p, ϵ) in E the edge from v to ϵ labeled by p and L is not a code.
    - for p in L
        - if vp is the prefix of a word of L then
            — if $vp \notin V$ then add vp in V
            — add (v, p, vp) in E the edge from v to vp labeled by p.
        - if vp is a word of L then add (v, p, ϵ) in E the edge from v to ϵ labeled by p and L is not a code.

Then we define the automaton $\mathcal{A} = (Q, \epsilon, T, \delta)$ by

$$\text{the set of states } Q = V$$
$$\text{the initial state } \epsilon$$
$$\text{the set of final states whish is } T =$$
$$\{\epsilon\} \text{ for finite words and T=V for infinite ones}$$
$$\text{and } \delta \text{ the function of transition defined from } E \text{ by for each } (u, l, v) \in E$$
$$\delta(u, l) = v$$

For infinite words, the recognition is done by the Büchi condition of acceptance of ω-languages[PP].

Proposition 1 *The automaton $\mathcal{A}$ recognizes :*

- *the set of words of L^+ with at least two factorizations, with $T = \{\epsilon\}$*
- *the set of words of L^ω with at least two factorizations, with $T = V$*

Proof. By the construction done by the algorithm, all words recognized by the automaton $\mathcal{A}$ have two factorizations.
Let us consider α such a word not recognized by $\mathcal{A}$. Let $v \notin V$ be a delay of one factorization on the other and $v' \in V$ it previous delay. There are only two cases:

 - $v \in \text{pref}(L)$ and $vv' \in L$ then $v \in V$.
 - $v \in \text{pref}(L)$, $v' \in \text{pref}(v)$ and $v'^{-1}v \in L$ then $v \in V$.

then $\mathcal{A}$ accept the word α.   □

This automaton can be non-deterministic and ambiguous (i.e one word can be accepted by two path in the automaton).

So the two following results hold :

Proposition 2 *Let L be a language, $G = (V, E)$ the associated graph. L is a code iff there is no cycle in the graph G that comes back to the vertex labeled by ϵ.*

Proof. If L is not a code then there is a finite word u with two factorizations and the final delay between the two factorizations is ϵ. The graph contains a cycle that comes back to the vertex labeled by ϵ.

Now, if the graph has a cycle path which begins by the vertex labeled by ϵ and comes back to this vertex, then there is a finite word with two factorizations and then, the language L is not a code.   □

Proposition 3 *Let L be a language, $G = (V, E)$ the associated graph. L is an ω-code iff there is no cycle in G.*

Proof. If there is a cycle then we can read an infinite number of label of edges which represent an infinite word with two factorizations, so the language L is not an ω-code.

If L is not an ω-code then there is at least one infinite word in L^ω with two factorizations on L. We can read an infinite sequence of labels l (of edges between two vertices). Sets V and E are finite because the language L is finite. Then to read an infinite sequence of l in the graph, we must reach more than one time some vertices and then there is a cycle in the graph.   □

Example 1. Let $L = \{a, abac, ba, ca\}$ a language. Let's consider languages L^* and L^ω.

 - The set of finite words with two factorizations is $a(baca)^+$
 - The set of infinite words with two factorizations is $(a(baca)^+)^*((abac)^\omega \cup (abaca)^\omega)$

The graph of the figure 2 represents all those words.

In this automaton we read only words with at least two factorizations and the label of a reaching vertex is the advance of one factorization on the other.

When we read a word with two factorizations, at the beginning the delay of one factorization on the other is ϵ.

If on one factorization we read a the delay is a:

So if we read ba, the factorization continues with aba and the delay is aba:

$$a \ b \ a \ldots$$

Now, reading c, the upper factorization begins with $abac$ and the delay is c:

$$a \ b \ a \ c \ldots$$

Then the lower factorization can continue with ca and the delay is a:

$$a \ b \ a \ c \ a \ldots$$

Here, if in the upper factorization we read a, we have a word of L^+ with two factorizations. But, we can continue by reading ba in the lower factorization. So we can read all the finite words with two factorizations. Those words of L^+ are represented by a finite path in the graph beginning and finishing by ϵ.
The infinite word with two factorizations is represented by the infinite path beginning by ϵ and infinitely crossing vertices labeled a, aba, c, and ϵ.

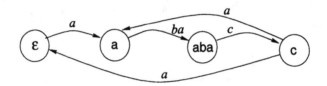

Fig. 2. The automaton for the language $\{a, abac, ba, ca\}$

4   The Software

The Implementation

The software is written in *Java 1.1*. We implement the automaton as a graph by adjacency list. We construct the graph using the algorithm. During the construction of the graph, we decide if the given language is a code, and an ω-code (if an edge come back to the vertex labeled by ϵ then the language is not a code, and if an edge come back to a vertex previously calculated then the language is not an ω-code.

The User Interface

The software has graphical interface. It needs a finite language as a string (*e.g.* "a,abac,ba,ca") which is given in a text field, and then draws the automaton and checks if the language is a code or an ω-code. There is an other text field to enter a word, also as a string (*e.g* "a(baca)∧2 "), this to decide if the word has more than one factorization on the given language.

References

[BP85]   J. Berstel and D. Perrin. *Theory of codes*. Academic Press, 1985.

[JLP96]   S. Julia, I. Litovsky, and B. Patrou. On codes, ω-codes and ω-generators. *IPL*, 60(1):1–5, 1996.

[Jul96]   S. Julia. *Sur les codes et les ω-codes générateurs de langages de mots infinis*. PhD thesis, Université de Nice - Sophia Antipolis, 1996.

[Kar85]   J. Karhumäki. A property of three-element codes. *TCS*, 41:215–222, 1985.

[MMP$^+$95] O. Matz, A. Miller, A. Potthoff, W. Thomas, and E. Valkema. Report on the program AMoRE. Technical report, Institut für Informatik und Praktische Mathematik, 1995. Software.

[PP]   D. Perrin and J.E. Pin. Mots infinis. (to be published) http://www.liafa.jussieu.fr/~jep/Resumes/MotsInfinis.html.

[SP53]   A.A. Sardinas and C.W. Paterson. A necessary and sufficent condition for the unique decomposition of coded messages. *IRE Internat. Conv. Rec.*, 8:104–108, 1953.

[VUMB97] J. Vöge, S. Ulbrand, O. Matz, and N. Buhrke. The automata theory package *omega*. In *proc. WIA '97, LNCS*, volume 1436. Springer, 1997.

Autographe: A Graphical Version of Automate*

Are Uppman and Jean-Marc Champarnaud

LIFAR
Université de Rouen, Faculté des Sciences et des Techniques
76821 Mont-Saint-Aignan Cedex, France
{uppman, champarnaud}@dir.univ-rouen.fr

Abstract. Autographe is a graphical version of Automate, a system for symbolic computation on automata and semigroups. Autographe is developed upon Egg, a graphical editor of graphs and a library of Java 1.1 classes aimed at the graphical treatment of oriented graphs, their vertices and edges and associated informations. Communication between the Automate server and the Egg editor is realized via sockets.

1   Introduction

Autographe is a system for symbolic computation on automata with graphical edition based upon a client/server-type collaboration between the two following programs :
   – Automate, a system for manipulating automata and semigroups developed by J.-M. Champarnaud and G. Hansel [4].
   – Egg, a graphical editor for graphs developed by A. Uppman [11].

2   The Automate Program

The Automate application lets you perform all the classical operations on regular expressions, automata and semigroups. It has been modified in order to play the role of a computation server for the Egg editor.

3   The Egg Editor

At the first, Egg (as "Editeur graphique de graphes") was developed as a programming exercice in C++ and Motif [12] and then in Java 1.0. Later on, Egg evolved into a library of Java 1.1 classes. This library offers classes that let you manage graphs through graphical and interactive views. Information may be attached to the vertices and the edges. Egg has actually been used to build a userfriendly interface to Automate. Communication between the two is done by sockets.

---

* This work is a contribution to the Automate software development project carried on by A.I.A. Working Group (Algorithmics and Implementation of Automata), L.I.F.A.R. Contact: {Champarnaud, Ziadi} @dir.univ-rouen.fr.

J.-M. Champarnaud, D. Maurel, D. Ziadi (Eds.): WIA'98, LNCS 1660, pp. 226–229, 1999.

3.1 Principles

If you define a graph as a set of vertices and a set of edges joining these vertices, then notions like:

- the position of vertices,
- the graphical form of vertices and edges,
- the information carried by edges and vertices

are absent.

The principles underlying the library Egg are then the following :

- the library should offer the basical classes and handle the most complex problems arising when you try to draw vertices and edges,
- inheritance should be used to adapt these classes to any type of information attached to vertices or edges,
- any visual representation of a graph (form, position) is arbitrary in the sense that it is one representation among many possible ones (thus several "views" on a single graph should be allowed).

3.2 Contents

The Egg library primarily tries to conform to the principles stated above. As a consequence data structures are distributed on several classes and often coded by java.util.Vector or java.util.Hashtable. In short : we do not privilege time or space optimisation.

Basical classes are Graph, Vertex and Edge. Their implementation is inspired by [3]. These classes don't care with any screen drawing. The graphical drawing is devoted to the Point and ArticulatedLine classes and their descendants. Among these classes, PointVertex and LineEdge respectively point to a vertex and to an edge (notice that the vertex and the edge are not aware of this).

Usually a PointVertex or a LineEdge is defined in an instance of the View class. Any number of views may be defined on the same graph. The view does not contain any information apart from form and position attributes. Any modification of the graph or its information (other than form and position) is thus transmitted to all of the views of the graph : thus the edition of the graph may be done in any view.

When a graph is given without any information on the positions of the vertices it's always a tedious task to position them by hand. Egg offers the Placer class to perform this task. Various placement algorithms are ready for use.

4 Functionalities of Autographe

As we have seen the user interface of the Autographe client-server system provides a graph editor built on the Egg library and a text editor for regular expressions at the client side, and the Automate program at the server side.

A typical Autographe use case is as follows. The user edits a regular expression and sends it to the server to get an automaton that recognizes the language of the expression. A standard representation of the automaton is presented by Autographe in a view. The user may display other non-standard views and edit graphically one or several views. The automaton as well as the views may be saved on disk.

Many other functionalities concerning edition of graphs (or views) are present. However, in this release many important functionalities are yet absent. For example, most of the functionalities of the Automate server are not available, but will be in next release.

We are also waiting for more user feedback to improve ergonomy at the client side.

5   Perspectives and Related Works

The graphical interface to the Automate command is in progress, offering the possibility of the graphical edition of an automaton. The graphical interface to the command Monoide is to come, offering the visualization of the results of this command. The class architecture of the Egg library will also easen the construction of a graphical interface for SEA^2 software which is made of the Maple packages AG [6], MULT [7], and TROP [8].

The editor of the FLAP [9] system allows to construct in a window a finite automaton, a stack automaton or a Turing machine. The FSGraph editor of the toolbox INTEX [10] allows to construct finite state transducers. The Egg editor makes available the same type of functionalities for finite state automata. In addition it allows to display an automaton from its transition table. This possibility is present in systems which provide animation for string matching algorithms as [5] or Padnon [2]. Egg uses a state placement algorithm which is similar to the one of the AMoRE system [1].

References

1. O. Matz, A. Miller, A. Potthoff, W. Thomas and E. Valkena, Report on the Program AMoRE, *Technical Report*, Institut für Informatik und Praktische Mathematik, Christian-Albrechts Universität, Kiel, 1995.
2. F. Bertault, Padnon : un Logiciel de Tracé Automatique de Graphes en Trois Dimensions, *Research Report 3130*, INRIA, 1997.
3. G. Booch, Software Components With Ada, Second Edition, *The Benjamin / Cummings Series in Ada and Software Engineering*, 1986.
4. J.-M. Champarnaud and G. Hansel, AUTOMATE: a Computing Package for Automata and Finite Semigroups, *J. Symbolic Computation* **12**, (1991) 197–220.

^{2} The development of the interface GTSEA (Graphic Tools for a Symbolic Environment of Automata) is supported by the PRC/GDR ALP (OMaMi group).

5. C. Charras and T. Lecroq, Exact String Matching Algorithms, *Report LIR*, Université de Rouen, 1997, URL : http://www.dir.univ-rouen.fr/~charras/string/.

6. P. Caron, AG : a Set of Maple Packages for Manipulating Automata and Semigroups, *Software–Practice & Experience*, 27(8), 863–884, 1997.

7. G. Duchamp, M. Flouret and E. Laugerotte, Operations over Automata with Multiplicities, Workshop on Implementing Automata, *WIA '98*, Rouen, 1988, to appear in Lecture Notes in Computer Science.

8. M. Kanta et D. Krob, TROP: a Maple Package for Tropical Automata, *Report LIAFA*, Université Paris 7, 1998, to appear.

9. S. Rodger, Integrating Hands-on Work into the Formal Languages Course via Tools and Programming, in: D. Raymond and D. Wood, eds., Proc. *WIA '96*, Lecture Notes in Computer Science, Vol. 1260 (1997) 132–148.

10. M. Silberztein, INTEX: an FST Toolbox, in: D. Wood and S. Yu, eds., Proc. *WIA '97*, Lecture Notes in Computer Science, Vol. 1436 (1988) 185–197.

11. A. Uppman, Egg: a Graphical Editor of Graphs, *Technical Report LIFAR*, Université de Rouen, 1998, to appear.

12. D. A. Young, Object-Oriented Programming with C++ and OSF/Motif, Second Edition, *Prentice Hall*, 1995.

INTEX 4.1 for Windows: A Walkthrough

Max Silberztein

GRELIS
Université de Franche-Comté
30 rue Megevand, 25000 Besancon
silberz@ladl.jussieu.fr

Abstract. INTEX is a linguistic development environment that allows users to build large-coverage finite state descriptions of natural languages and apply them to large texts in real time. INTEX represents texts, grammars and dictionaries by Finite State Transducers.

1 Introduction

INTEX is a linguistic development environment that allows users to build large-coverage finite state descriptions of natural languages and apply them to large texts (several dozen million words) in real time.

One important aspect of INTEX is that grammars, dictionaries, as well as texts are all represented by Finite State Transducers (FSTs). Therefore, all the operations users perform via the graphical interface are translated into elementary operations on FSTs. For instance, applying a set of dictionaries to a text is performed by constructing a union of the dictionaries' FSTs, and then applying the resulting FST to the text FST; removing lexical ambiguities in the text is performed by computing the intersection between a grammar's FST and the text's FST, etc.

INTEX includes a hundred tools in 3 areas:

- tools that process texts: indexing a text, checking its format, segmenting it into text units (e.g. sentences), normalizing special words (e.g. elisions or contractions) or spelling variants, indexing sequences that match a FST, constructing the resulting highlighted text or concordance, extracting text units that contain some (or no) matching sequences, applying statistical tools to study the vocabulary, the frequency, the coverage and the evolution of the matching sequences, displaying the FST that represents the text, etc. (these tools are available under the Menu Text);

- tools that process dictionaries: checking the spelling of the entries of a dictionary, the correctness of its codes, as well as its format described by a FSA, sorting a dictionary, inflecting a DELAS-type dictionary, compacting a DELAF-type dictionary into a FST (Menu Dictionaries);

J. M. Ch...   d. D. M... d. D. Ziadi (Eds.): WIA'98, LNCS 1660, pp. 230–243, 1999.

- tools that process grammars: the graphical editor, compiling large sets of graphs into FSTs, formatting graphs, etc. (Menu FSGraph)

2   Applying INTEX FSTs to Texts

FSTs may be applied to texts in order to modify them in three modes:

- in REPLACE mode, matching sequences are replaced with the corresponding FST output; this feature is similar to the Search & Replace operations of the regular word processors;

- in MERGE mode, FST outputs corresponding to matching sequences are inserted into the text; each inserted output sequence is inserted before the corresponding input sequence;

- in DISAMBIGUATING mode, FST outputs consist of sequences of lexical constraints that must be applied to the corresponding lexical entries of the FST input. The result is a partially tagged text;

When applying FSTs to modify a text, it is important to set some priority Rules.

2.1   FSTs Are Applied from Left to Right

For instance, consider the text (1) and the FST of Figure 1. if we apply the FST of Figure 1 to the text (1), we get the results (2) or (3).

$$z\ a\ b\ c\ d\ e\ z \tag{1}$$

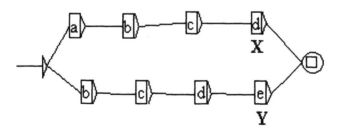

Fig. 1.

in REPLACE mode:

$$z \, X \, e \, z \qquad\qquad (2)$$

in MERGE mode:

$$z \, a \, b \, c \, X \, d \, e \, z \qquad\qquad (3)$$

In other words, the text was modified before the sequence $b \, c \, d \, e$ was recognized.

There is a special case for this rule: sequences that start with the special symbol $< \wedge >$ (beginning of text unit). In INTEX, text units are either:

- lines or paragraphs delimited by the character NEWLINE,

- concordance entries, i.e. 3-column lines, where the center column contains previously indexed sequences, and the left and right columns are length-fixed contexts of the utterances;

- any text unit delimited by the special mark {S}; usually, this mark has been introduced in the text with a special FST that identifies sentences.

The special symbol $< \wedge >$ always have priority over sequences that do not start in this way; for instance, if we apply the FST of Figure 2 to a text, the sequences $a \, b$ that occur at the beginning of text units will be replaced with Y, while sequences that occur elsewhere in the text unit will be replaced with X. In other words, at the beginning of text units, the matching sequence $< \wedge > \, a \, b$ has priority over the matching sequence $a \, b$.

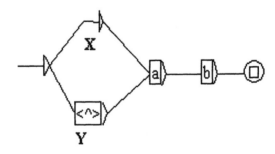

Fig. 2.

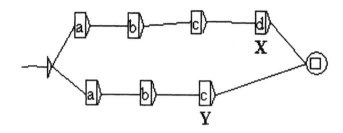

Fig. 3.

2.2   Longest Matches Have Priority over Shorter Ones

For instance, we apply the FST of Figure 3 to the text (1).

We get the same results (2) and (3) because the matching sequence *a b c d* has priority over the shorter one *a b c*. This priority rule is natural for the disambiguation process, where longest matching sequences are better disambiguated than shorter ones; this allows also linguists to write default disambiguation rules, such as in the FST of Figure 4

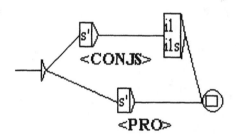

Fig. 4.

If *s'* is followed by *il* or *ils*, it is a conjunction (=*si*); otherwise, it is a pronoun (=*se*). In the first case, the matching sequence *s'il* has priority over the other matching sequence *s'*.

2.3   Ambiguous FSTs Produce an Undefined Result

An ambiguous FST is a FST that associates one matching sequence with more than one output sequence.

Note that this problem only occurs when one applies an ambiguous FST to a text which is represented in a linear way (e.g. ASCII text). When the text is itself represented by a FST, applying ambiguous FSTs is perfectly valid and it will lead to the creation of parallel paths in the text FST.

2.4 ϵ-FSTs Cannot Be Applied to Texts

An ϵ-FSTs is a FST that recognizes the empty string.

3 A Walkthrough

Now I will comment on several screen shots that correspond to important stages of text parsing:

1. Processing an ASCII-ISO 'raw' text in order to prepare it for further linguistic analysis. Generally, this preprocessing stage consists of three application of FSTs to the text:
   - a FST used in MERGE mode to insert text unit delimiters (usually sentence delimiters),
   - a lexical FST used in REPLACE mode to tag unambiguous compound words,
   - a FST used in REPLACE mode to solve elisions and contractions (such as *can't* for *can not*).

2. Identifying words in the text consists of applying selected dictionaries and morphological grammars to the text. Linguistic words are either simple words (lexical entries that are sequences of letters) or compound words (lexical entries that contain more than one simple word). Dictionaries are in the form of DELAFs, that is, their entries are inflected forms (e.g. *eaten*), and are associated with both a corresponding lemma (e.g. *eat*) and some kind of information (e.g. verb, participle). Morphological grammars are graphs that identify families of (potentially infinite) form variants and produce both a corresponding canonical form (a lemma, or a standard keyword for orthographic variants or synonymous expressions) and some kind of linguistic information.

3. Generally, consultation of the dictionaries produces more than one lexical entry for each form in the text. The next step is to remove these ambiguities by applying local grammars in the form of FSTs that associate certain matching ambiguous sequences of forms with some lexical constraints. These constraints are in turn applied to the forms during consultation of the dictionaries in order to destroy irrelevant lexical hypotheses.

4. Applying a FST to a text to locate matching sequences. The resulting index can be used to highlight the text (all matching sequences are colored and underlined), to extract all the text units that contain (or do not contain any) matching sequences, or to display all the matching sequences in context (concordance). Matching sequences can be also studied with several statistical tools.

5. Constructing the FSTs that represent each sentence of the text; in this FST, inputs are lemma, and outputs are the corresponding lexical information. Ambiguities between simple words, or between a compound word and the corresponding sequence of simple words are represented by parallel paths. This FST is meant to be the input of the INTEX syntactic parser.

3.1 Preprocessing the Text

An ASCII text is loaded; the FST Sentence is applied to the text in MERGE mode to insert the text unit delimiter {S} between sentences (Figure 5).

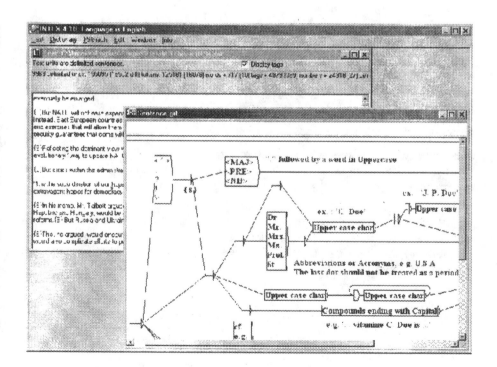

Fig. 5.

In this notation, states are hidden (they can be imagined as being at the left side of every box). Inputs are written inside boxes; outputs are written below boxes. <E> stands for the empty string, <MAJ> stands for any word in uppercase letters, <NB> stands for any sequence of digits. Gray nodes are names of embedded FSTs; for instance, *Upper case char* is the name of an embedded FST that identifies the 26 letters A...Z.

The path at the top of the graph matches sequences of a period, an interrogation mark or an exclamation mark followed by a word in uppercase; the symbol {S} is then inserted between the period and the word. Below is a path that matches sequences such as *Prof. C. Doe*. For such sequences, no sentence delimiter is inserted.

FSTs built in INTEX are fully editable and customizable; there is a different FST used to identify sentences for each language.

3.2   Simple and Compound Words

Lexical FSTs and dictionaries are applied to identify simple and compound Words (Figure 6)

After the preprocessing stage, users select dictionaries and FSTs used for the recognition of simple and compound words in the text. Dictionaries are in the form of a DELAF dictionary, that is, they contain entries as inflected forms (e.g. *ate*) and associate them with a lemma (*eat*), as well as some kind of linguistic information such as a syntactic category (*V*) and some inflectional information (*Preterit*).

FSTs must be used when lexical entries are infinite in number (e.g. numerical determiners such as *two hundred and fifty six*); they are also conducive to putting together families of orthographic variants, morphological derivations, or synonymous expressions.

Tokens of the text are either sequences of letters (words), sequences of digits (numbers), tags (linguistic information written between curly brackets) or delimiters (all the characters that are neither letters nor digits). Tokens are sorted by decreasing frequency.

Users select dictionaries and FSTs to be applied to the text. Users can edit dictionaries and FSTs and add their own data. Dictionaries and FSTs are associated with a 3-level priority system that allows hiding or imposing some kind of lexical information.

The result of the consultation of all selected dictionaries is presented below, in four windows (Figure 7).

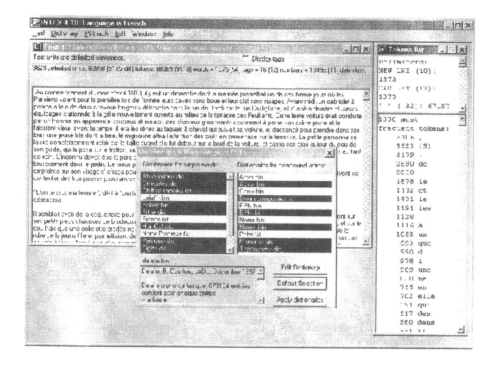

Fig. 6.

- all the simple words that correspond to one or more lexical entries are presented in the top left window, which is associated with corresponding lemma and linguistic information. Ambiguous forms are presented on several lines;

- all the simple words that have not been found in any selected dictionary, nor matched any lexical FST are displayed in the right top window. Usually, these forms are either spelling errors, proper names, or words specific to the domain;

- the compound words are displayed in the lower two windows. Certain forms are a priori ambiguous, such as *red tape* (either the adjective followed by the noun, or the compound noun). Other compounds are not ambiguous, such as *best-seller* or *acquired immune deficiency syndrom*.

Users can edit these resulting dictionaries in order to remove standard lexical ambiguities that by chance do not occur in the specific text, or to move words from the unknown words window to the simple words window, without modifying the general dictionaries of the system.

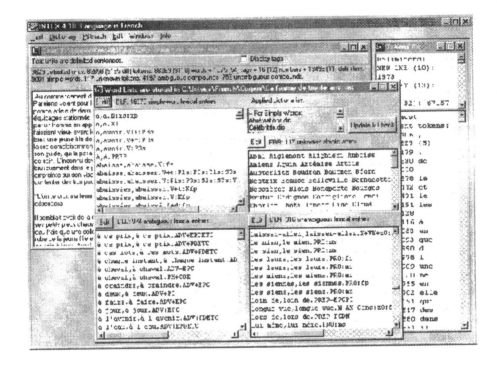

Fig. 7.

3.3 Disambiguation of Grammatical Words

When one Applies local grammars to disambiguate grammatical words in a French Text, each local grammar is an FST that recognizes certain text sequences (e.g. Figure 8 *il le la donne*) and then applies the corresponding lexical constraints (e.g. <PRO> <PRO> <PRO> <V:3s>). All the lexical hypotheses that do not match the lexical constraints are then deleted.

The result displayed above is the tagged text, i.e. the full text in which all the disambiguated forms have been replaced by the corresponding tag, which appears between curly brackets. For instance, the original text starts with (4).

C'est en Egypte, vers la fin de...   (4)

After the disambiguation process, the text becomes (5)

{C',ce.PRO+PPV:ms} {est,être.V:P3s} en Egypte,

vers {la,le.DET:fs} fin {de,.PREP}...   (5)

{*C',ce.PRO+PPV:ms*} stands for form *C'*, corresponding to lemma *ce*, pronoun, preverbal particle, masculine singular; {*est,être.V:P3s*} stands for form *est*, corresponding to lemma *être*, verb, present, 3rd person singular, etc.

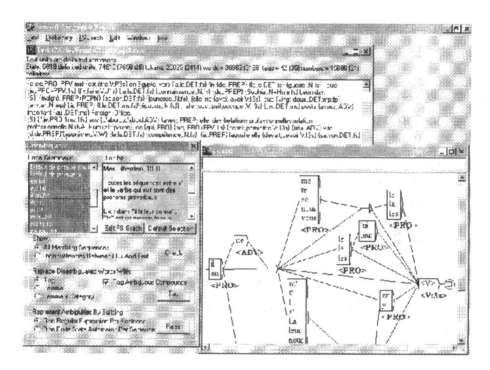

Fig. 8.

In a tagged text, all the forms that are ambiguous are left unmodified. It is also possible to display the remaining ambiguities by displaying either a regular expression or the finite state transducer of the text.

3.4   Indexing an FST

- It is possible to index sequences that match a regular expression, such as:
  <be:P> (<ADV> + <E>) going to <V:W> + (will + shall) <V:W>

  all the forms of the word "be" conjugated in the present tense, followed by an optional adverb, followed by "going to", followed by any verb in the infinitive or the word "will" or "shall" followed by a verb in the infinitive. Any user- defined code that appears in a dictionary can automatically be used in a symbol (between angles);

- one can also locate all the entries of a compound word dictionary. In this way, technical terms can be indexed and extracted;

– one can also use the graphical editor to construct grammars; these graphs can be applied to text (Figure 9).

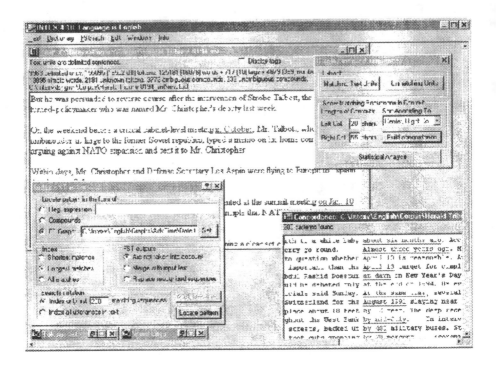

Fig. 9.

Above, we have applied a FST that recognizes certain complements of dates (over 8000 states in over 100 graphs) to the text. Matching sequences are highlighted in the top left window; the corresponding concordance is displayed in the lower right-hand window.

We can extract all the sentences that contain one or more matches (or no match), from the text; several statistical functions can be applied to study the density of matches in the text, the evolution of matching sequences throughout the text, the *vocabulary* of matching sequences (i.e. the set of all different matching sequences that match the FST), etc.

3.5   Text Representation

Internally, each text sentence is represented by a FST. For instance, the Text 6 is represented on Figure 10 with all the ambiguities that remain after having applied INTEX default local grammars.

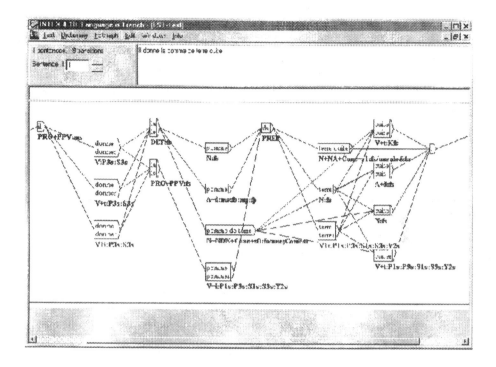

Fig. 10.

Il donne la pomme de terre cuite   (6)

Thanks to several local grammars, only the 3rd person singular of the verb *donner* is displayed; the determiner *la* can be followed by the noun or adjective *pomme*, or the noun *pomme de terre*, while the pronoun *la* can only be followed by the verb *pommer*; the form *de* has also been disambiguated.

Note that at this stage of analysis, the computer cannot decide if the correct interpretation is:

- He eats a clay apple, (compound noun *terre cuite* = clay)
- He eats a cooked potato (compound noun *pomme de terre* = potato)

4   INTEX as a Research Tool

Over 30 research centers are presently using INTEX as a research tool in various domains: computational linguistics, corpus-based linguistics, information retrieval, terminology, literature studies, and teaching of a second-language.

Large-coverage INTEX descriptions have already been built for Bulgarian (Prof. Elena Paskaleva, University of Sofia), English, French and Spanish (Prof. Maurice Gross, University of Paris 7), German (Prof. Franz Guenthner, University Maximilian, Munich), Greek (Prof. Ana Symeonides, University of Salonique), Italian (Prof. Annibale Elia, University of Salerne), Polish (Prof. Zygmunt Vetulani, University of Gdansk), Portuguese (Prof. Elisabete Ranchodd, University of Lisbon), Russian (Alex Kolesov, Academy of Sciences, Moscow). INTEX Dictionaries are being constructed for Korean (Jeesun Nam, University of Seoul) and Old French (Prof. Hava Bat-Zeev, University of Tel Aviv).

See [Silberztein 1996a] for the proceedings of the first INTEX users' Workshop. 60 participants came from 10 countries; 15 papers demonstrated the various applications of the system.

Two European Projects have used the INTEX environment:

- to build a computer aided translation workstation (EUROLANG used INTEX technology and linguistic data to automatically identify terms in texts);

- to build large-coverage descriptions of Bulgarian, German, Polish and Russian (BILEDITA project).

5   INTEX as an Education Tool

Describing the linguistic data included in INTEX (dictionaries for simple words, description of the morphology, dictionaries for compounds, dictionaries for frozen expressions, local grammars used to tag and disambiguate texts) as well as the technology used to handle this data (Finite State Automata and Transducers) corresponds to a full year course for graduate students. There are a plethora of projects that remain to be proposed to students, as the description of natural languages is nowhere near complete!

INTEX is also used to teach French as a second-language; it allows teachers to ask students to locate morpho-syntactic patterns in wide texts (such as 5 years of the newspaper *Le Monde*), to build local grammars for semi-frozen expressions (e.g. how to express a date in French) and to 'play' (edit, correct, apply and test) with some linguistic descriptions.

References

1. Silberztein M. (1998), *Presentation, Multimedia tutorials, INTEX 3.4 reference manual*, www.ladl.jussieu.fr/INTEX.
2. Silberztein M. (1993), *Dictionnaires électroniques et analyse automatique de textes*, Paris: Masson.

3. Silberztein M. (1994), *INTEX: A Corpus Processing System*, COLING'94. Kyoto.
4. Silberztein M. (1997), *The Lexical Analysis of Natural Languages*, in *Finite State Language Processing* E. Roche and Y. Schabes eds. Cambridge: The MIT Press.
5. Silberztein M. ed. (1996), *Proceedings of the First INTEX users workshop*, Paris: LADL, Université Paris 7.

Author Index

Lecture Notes in Computer Science

For information about Vols. 1–1632
please contact your bookseller or Springer-Verlag